ANALYSIS

This self-contained text, suitable for advanced undergraduates, provides an extensive introduction to mathematical analysis, from the fundamentals to more advanced material. It begins with the properties of the real numbers and continues with a rigorous treatment of sequences, series, metric spaces, and calculus in one variable.

Further subjects include Lebesgue measure and integration on the line, Fourier analysis, and differential equations. In addition to this core material, the book includes a number of interesting applications of the subject matter to areas both within and outside of the field of mathematics. The aim throughout is to strike a balance between being too austere or too sketchy, and being so detailed as to obscure the essential ideas. A large number of examples and nearly 500 exercises allow the reader to test understanding and practice mathematical exposition, and they provide a window into further topics.

Richard Beals is James E. English Professor of Mathematics at Yale University. He has also served as a professor at the University of Chicago and as a visiting professor at the University of Paris, Orsay. He is the author of more than 100 research papers and monographs in partial differential equations, differential equations, functional analysis, inverse problems, mathematical physics, mathematical psychology, and mathematical economics.

ANALYSIS

An Introduction

RICHARD BEALS
Yale University

PUBLISHED BY THE PRESS SYNDICATE OF THE UNIVERSITY OF CAMBRIDGE
The Pitt Building, Trumpington Street, Cambridge, United Kingdom

CAMBRIDGE UNIVERSITY PRESS
The Edinburgh Building, Cambridge CB2 2RU, UK
40 West 20th Street, New York, NY 10011-4211, USA
477 Williamstown Road, Port Melbourne, VIC 3207, Australia
Ruiz de Alarcón 13, 28014 Madrid, Spain
Dock House, The Waterfront, Cape Town 8001, South Africa

http://www.cambridge.org

© Richard Beals 2004

This book is in copyright. Subject to statutory exception
and to the provisions of relevant collective licensing agreements,
no reproduction of any part may take place without
the written permission of Cambridge University Press.

First published 2004

Printed in the United States of America

Typeface Times 11/14 pt.   *System* LATEX 2_ε [TB]

A catalog record for this book is available from the British Library.

Library of Congress Cataloging in Publication Data

Beals, Richard, 1938–
Analysis : an introduction / Richard Beals.
p. cm.
Includes bibliographical references and index.
ISBN 0-521-84072-4 – ISBN 0-521-60047-2 (pbk.)
1. Mathematical analysis.   I. Title.
QA300.B4124   2004
515–dc22   2004040687

ISBN 0 521 84072 4 hardback
ISBN 0 521 60047 2 paperback

Contents

	Preface	*page* ix
1	Introduction	1
	1A. Notation and Motivation	1
	1B*. The Algebra of Various Number Systems	5
	1C*. The Line and Cuts	9
	1D. Proofs, Generalizations, Abstractions, and Purposes	12
2	The Real and Complex Numbers	15
	2A. The Real Numbers	15
	2B*. Decimal and Other Expansions; Countability	21
	2C*. Algebraic and Transcendental Numbers	24
	2D. The Complex Numbers	26
3	Real and Complex Sequences	30
	3A. Boundedness and Convergence	30
	3B. Upper and Lower Limits	33
	3C. The Cauchy Criterion	35
	3D. Algebraic Properties of Limits	37
	3E. Subsequences	39
	3F. The Extended Reals and Convergence to $\pm\infty$	40
	3G. Sizes of Things: The Logarithm	42
	Additional Exercises for Chapter 3	43
4	Series	45
	4A. Convergence and Absolute Convergence	45
	4B. Tests for (Absolute) Convergence	48
	4C*. Conditional Convergence	54
	4D*. Euler's Constant and Summation	57
	4E*. Conditional Convergence: Summation by Parts	58
	Additional Exercises for Chapter 4	59

5	Power Series	61
	5A. Power Series, Radius of Convergence	61
	5B. Differentiation of Power Series	63
	5C. Products and the Exponential Function	66
	5D*. Abel's Theorem and Summation	70
6	Metric Spaces	73
	6A. Metrics	73
	6B. Interior Points, Limit Points, Open and Closed Sets	75
	6C. Coverings and Compactness	79
	6D. Sequences, Completeness, Sequential Compactness	81
	6E*. The Cantor Set	84
7	Continuous Functions	86
	7A. Definitions and General Properties	86
	7B. Real- and Complex-Valued Functions	90
	7C. The Space $C(I)$	91
	7D*. Proof of the Weierstrass Polynomial Approximation Theorem	95
8	Calculus	99
	8A. Differential Calculus	99
	8B. Inverse Functions	105
	8C. Integral Calculus	107
	8D. Riemann Sums	112
	8E*. Two Versions of Taylor's Theorem	113
	Additional Exercises for Chapter 8	116
9	Some Special Functions	119
	9A. The Complex Exponential Function and Related Functions	119
	9B*. The Fundamental Theorem of Algebra	124
	9C*. Infinite Products and Euler's Formula for Sine	125
10	Lebesgue Measure on the Line	131
	10A. Introduction	131
	10B. Outer Measure	133
	10C. Measurable Sets	136
	10D. Fundamental Properties of Measurable Sets	139
	10E*. A Nonmeasurable Set	142
11	Lebesgue Integration on the Line	144
	11A. Measurable Functions	144
	11B*. Two Examples	148
	11C. Integration: Simple Functions	149
	11D. Integration: Measurable Functions	151
	11E. Convergence Theorems	155

12	Function Spaces	158
	12A. Null Sets and the Notion of "Almost Everywhere"	158
	12B*. Riemann Integration and Lebesgue Integration	159
	12C. The Space L^1	162
	12D. The Space L^2	166
	12E*. Differentiating the Integral	168
	Additional Exercises for Chapter 12	172
13	Fourier Series	173
	13A. Periodic Functions and Fourier Expansions	173
	13B. Fourier Coefficients of Integrable and Square-Integrable Periodic Functions	176
	13C. Dirichlet's Theorem	180
	13D. Fejér's Theorem	184
	13E. The Weierstrass Approximation Theorem	187
	13F. L^2-Periodic Functions: The Riesz-Fischer Theorem	189
	13G. More Convergence	192
	13H*. Convolution	195
14*	Applications of Fourier Series	197
	14A*. The Gibbs Phenomenon	197
	14B*. A Continuous, Nowhere Differentiable Function	199
	14C*. The Isoperimetric Inequality	200
	14D*. Weyl's Equidistribution Theorem	202
	14E*. Strings	203
	14F*. Woodwinds	207
	14G*. Signals and the Fast Fourier Transform	209
	14H*. The Fourier Integral	211
	14I*. Position, Momentum, and the Uncertainty Principle	215
15	Ordinary Differential Equations	218
	15A. Introduction	218
	15B. Homogeneous Linear Equations	219
	15C. Constant Coefficient First-Order Systems	223
	15D. Nonuniqueness and Existence	227
	15E. Existence and Uniqueness	230
	15F. Linear Equations and Systems, Revisited	234
	Appendix: The Banach-Tarski Paradox	237
	Hints for Some Exercises	241
	Notation Index	255
	General Index	257

Preface

This text contains material for a two- or three-semester undergraduate course. The aim is to sketch the logical and mathematical underpinnings of the theory of series and one-variable calculus, develop that theory rigorously, and pursue some of its refinements and applications in the direction of measure theory, Fourier series, and differential equations.

A good working knowledge of calculus is assumed. Some familiarity with vector spaces and linear transformations is desirable but, for most topics, is not indispensable.

The unstarred sections are the core of the course. They are largely independent of the starred sections. The starred sections, on the other hand, contain some of the most interesting material.

Solving problems is an essential part of learning mathematics. Hints are given at the end for most of the exercises, but a hint should be consulted only after a real effort has been made to solve the problem.

I am grateful to various colleagues, students, friends, and family members for comments on, and corrections to, various versions of the notes that preceded this book. Walter Craig enlightened me about the difference between clarinets and oboes, and the consequences of that difference. Eric Belsley provided numerous corrections to the first version of the notes for Chapters 1–9. Other helpful comments and corrections are due to Stephen Miller, Diana Beals-Reid, and Katharine Beals. Any new or remaining mistakes are my responsibility.

I had the privilege of first encountering many of these topics in a course taught by Shizuo Kakutani, to whom this book is respectfully dedicated.

1
Introduction

The properties of the real numbers are the basis for the careful development of the topics of analysis. The purpose of this chapter is to engage in a preliminary and rather informal discussion of these properties and to sketch a construction that justifies assuming that the properties are satisfied. Along the way we introduce standard notation for various sets of numbers.

1A. Notation and Motivation

First, we use $\mathbb{N}$ to denote the set of *natural numbers* or *positive integers*:

$$\mathbb{N} = \{1, 2, 3, 4, \ldots\}.$$

In this set there are two basic algebraic operations, *addition* and *multiplication*. Each of these operations assigns to a pair of positive integers p, q an integer, respectively, the *sum* $p + q$ and the *product* $p \cdot q$ or simply pq. Further operations, such as powers, may be defined from these. There are then many facts, such as

$$1+2 = 3, \quad 1+2+4 = 7, \quad 1+2+4+8 = 15, \quad 1+2+4+8+16 = 31, \quad \ldots.$$

More interesting, from a mathematical point of view, are general statements, like

$$1 + 2 + 2^2 + 2^3 + \cdots + 2^n = 2^{n+1} - 1, \quad \text{all } n \in \mathbb{N}. \tag{1}$$

Within $\mathbb{N}$, there is also an *order relation*, denoted $<$, defined as follows. If m and n are elements of $\mathbb{N}$, then $m < n$ if and only if there is $p \in \mathbb{N}$ such that $m + p = n$. If so, we also write $n > m$. It is easy to convince oneself that this has the properties that *define* an "order relation" – given elements m, n of $\mathbb{N}$, exactly one of the following is true:

$$m < n, \quad \text{or} \quad n < m, \quad \text{or} \quad m = n. \tag{2}$$

Moreover, the relation is transitive:

$$m < n, \quad n < p \implies m < p. \tag{3}$$

(The one-sided arrow $\implies$ means "implies.")

Implicit in this discussion is the following fact: Given positive integers p and q, the equation $p + r = q$ does not generally have a solution r in $\mathbb{N}$; the necessary and sufficient condition is that $p < q$. Of course one can get around this difficulty by introducing $\mathbb{Z}$, the set

$$\mathbb{Z} = \{0, 1, -1, 2, -2, 3, -3, \ldots\}$$

of *integers*. The operations of addition and multiplication extend to this larger set, as does the order relation. Within this larger set one can make new statements, such as

$$1 - 2 + 2^2 - 2^3 + 2^4 - \cdots + (-2)^n = \frac{1 - (-2)^{n+1}}{3}. \tag{4}$$

The left side of this equation is clearly an integer, so the right side must also be an integer, despite the fact that not every integer is divisible by 3.

A more general way to put the statement about divisibility is this: Given integers p and q, the equation $qr = p$ does not generally have a solution r in the set $\mathbb{Z}$. To remedy this we must enlarge our set once more and go to $\mathbb{Q}$, the set

$$\mathbb{Q} = \{p/q : p \in \mathbb{Z}, q \in \mathbb{N}\}$$

of *rational numbers*. The operations of addition and multiplication extend to the larger set, as does the order relation. Here we may make a statement that generalizes (1)–(3):

$$1 + r + r^2 + r^3 + \cdots + r^n = \frac{1 - r^{n+1}}{1 - r} \quad \text{if } r \in \mathbb{Q} \text{ and } r \neq 1. \tag{5}$$

The identities (1)–(5) are purely algebraic. The last one leads to a kind of statement that has a different character. Suppose that r is "small": Specifically, suppose that $|r| < 1$. Then successive powers of r get smaller and smaller, so that one might be tempted to write

$$1 + r + r^2 + r^3 + \cdots = \frac{1}{1 - r}, \quad \text{if } r \in \mathbb{Q} \text{ and } |r| < 1. \tag{6}$$

Here the ellipsis $\cdots$ means that the addition on the left is imagined to be carried out for *all* powers of r, that is, there are infinitely many summands. The reader may or may not feel that it is clear what the left side means and why it is equal to the right side; these points will be discussed in much detail in Chapter 4.

1A. Notation and Motivation

Consider two more examples of statements like (6):

$$1 - \frac{1}{2} + \frac{1}{3} - \frac{1}{4} + \frac{1}{5} - \frac{1}{6} + \frac{1}{7} - \cdots = s_1; \tag{7}$$

$$1 + \frac{1}{3} - \frac{1}{2} + \frac{1}{5} + \frac{1}{7} - \frac{1}{4} + \frac{1}{9} + \frac{1}{11} - \frac{1}{6} + \cdots = s_2. \tag{8}$$

Note that the second (formal) sum has exactly the same summands as the first, except that they are written in a different order. We know that addition is associative and commutative, so it would seem that if the sums mean anything, then clearly $s_1 = s_2$. Now group the terms in (7):

$$s_1 = 1 - \left(\frac{1}{2} - \frac{1}{3}\right) - \left(\frac{1}{4} - \frac{1}{5}\right) - \left(\frac{1}{6} - \frac{1}{7}\right) - \cdots$$

$$= 1 - \frac{1}{2 \cdot 3} - \frac{1}{4 \cdot 5} - \frac{1}{6 \cdot 7} - \cdots. \tag{9}$$

Each expression in parentheses is positive, so we should have $s_1 < 1$. Similarly, in (8),

$$s_2 = 1 + \left(\frac{1}{3} - \frac{1}{2} + \frac{1}{5}\right) + \left(\frac{1}{7} - \frac{1}{4} + \frac{1}{9}\right) + \left(\frac{1}{11} - \frac{1}{6} + \frac{1}{13}\right) + \cdots$$

$$= 1 + \frac{1}{2 \cdot 3 \cdot 5} + \frac{1}{4 \cdot 7 \cdot 9} + \frac{1}{6 \cdot 11 \cdot 13} + \cdots. \tag{10}$$

Each expression in parentheses is positive, so we should have $s_2 > 1$. It is tempting to conclude that *either* the processes we are describing do not make sense *or* that there is some subtle flaw in the argument that purports to show that $s_1 \neq s_2$. However, the processes *do* make sense, and there is no flaw in the argument. In fact, in Chapter 4, Section D, we will show how to prove that

$$1 - \frac{1}{2} + \frac{1}{3} - \frac{1}{4} + \frac{1}{5} - \frac{1}{6} + \frac{1}{7} - \cdots = \log 2;$$

$$1 + \frac{1}{3} - \frac{1}{2} + \frac{1}{5} + \frac{1}{7} - \frac{1}{4} + \frac{1}{9} + \cdots = \frac{3}{2} \log 2,$$

where log 2 means the natural logarithm. In Chapter 5, Section D, we will show how to obtain different proofs of these identities. (We also present an argument for the "identity"

$$1 + 2 + 3 + 4 + 5 + 6 + 7 + \cdots = -\frac{1}{12},$$

but this last should not be taken too seriously.)

4 *Introduction*

There are a number of points to be made in this connection:

- Care must be taken with infinite repetition of algebraic operations.
- When care is taken, the results may be paradoxical but they are consistent, and often important.
- The addition of rational numbers can lead to an irrational sum, when it is extended to the case of infinitely many summands. Here is another example, also proved later:

$$1 + \frac{1}{4} + \frac{1}{9} + \frac{1}{16} + \frac{1}{25} + \frac{1}{36} + \cdots = \frac{\pi^2}{6}. \tag{11}$$

Let us pause to examine (11). Without worrying, at the moment, about the equality between the left and right sides, consider how one might conclude that the left side should have some meaning. The sequence of rational numbers

$$r_1 = 1, \qquad r_2 = 1 + \frac{1}{4}, \qquad r_3 = 1 + \frac{1}{4} + \frac{1}{9}, \qquad r_4 = 1 + \frac{1}{4} + \frac{1}{9} + \frac{1}{16}, \qquad \ldots$$

is *increasing*: $r_1 < r_2 < r_3 < \ldots$. We show later that this sequence is *bounded above*; in fact, $r_n < 2$ for every n. There is a standard representation of rationals as points on a line. We might expect geometrically that there is a unique point on the line with the property that, as n increases, the rationals r_n come arbitrarily close to this point. [Warning: Statements like "come arbitrarily close" need, eventually, to be made precise.] Then the left-hand side of (11) should be taken to mean the number that corresponds to this point. Thus, to be sure that things like the left side of (11) have a meaning, we want to be sure that *any bounded, increasing sequence of numbers has a limit*. (This is one version of what can be called the "no-gap" property of the real numbers. Starting in Chapter 2 we will take as basic a different, but equivalent, version, the "Least Upper Bound Property." See Figure 1.)

Another example of a bounded, increasing sequence is

$$3, \quad 3.1, \quad 3.14, \quad 3.141, \quad 3.1415, \quad 3.14159, \quad \ldots.$$

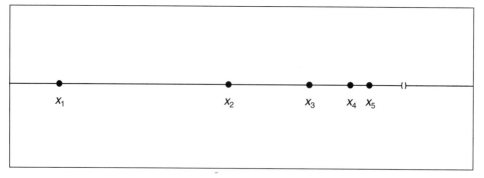

Figure 1. Heading for a gap.

Assuming that this sequence is headed where we expect, the limit π is known not to be a rational number. What justification do we have for asserting the existence of such numbers, and for thinking that we can add and multiply them in the usual ways, with the usual rules, such as $(x + y) + z = x + (y + z)$, without encountering a contradiction? These questions will be discussed in the next two sections.

Exercises

1. Prove the identity (1) by induction on n.
2. Prove by induction that the numerator in the right side of (4) is always divisible by 3.
3. (a) Prove the identity (5) by induction.
   (b) Give a second proof of the identity (5).
4. Derive and prove a general form for the expressions in parentheses in the sum (9), thus verifying that these expressions are positive.
5. Derive and prove a general form for the expressions in parentheses in the sum (10), thus verifying that these expressions are positive.

1B*. The Algebra of Various Number Systems

We begin by examining the "usual rules." The basic properties of addition and multiplication in $\mathbb{N}$ can be summarized in the following axioms (statements of properties). It is understood for the moment that m, n, and p denote arbitrary elements of $\mathbb{N}$.

A1: Associativity of addition. $(m + n) + p = m + (n + p)$.

A2: Commutativity of addition. $m + n = n + m$.

M1: Associativity of multiplication. $(mn)p = m(np)$.

M2: Commutativity of multiplication. $mn = nm$.

D: Distributive law(s). $m(n + p) = mn + mp$; $(m + n)p = mp + np$.

[Note that either part of D follows from the other part, together with M2.]

The order relation in $\mathbb{N}$ has the defining characteristics of an order relation. Again m and n denote arbitrary elements of $\mathbb{N}$,

O1: Trichotomy. *Exactly one of the following is true:* $m < n$, $n < m$, *or* $m = n$.

O2: Transitivity. *If* $m < n$ *and* $n < p$, *then* $m < p$.

The order relation has connections with addition and with multiplication:

$$m < m + n; \quad m < n \Rightarrow mp < np, \quad \text{all } m, n, p \in \mathbb{N}. \tag{12}$$

We shall take the positive integers and these properties for granted. One can then *construct* the set of all integers as follows. The *ingredients* are all the formal expressions $m - n$, where m and n are positive integers. This formal expression can be thought of as representing the "solution" x of the equation $n + x = m$. We do not want to consider these as all representing different things (consider $1 - 1$ and $2 - 2$), so we *identify* the expressions $m - n$ and $m' - n'$ under a certain condition:

$$m - n \equiv m' - n' \quad \text{if } m + n' = n + m'. \tag{13}$$

The set $\mathbb{Z}$ may be thought of, for now, as the set of such expressions, subject to the "identification" rule (13).

Addition and multiplication of these expressions are *defined* by

correction $\rightarrow$
$$(m - n) + (p - q) = (m + p) - (n + q); \tag{14}$$
$$(m - n) \cdot (p - q) = (mp + nq) - (mq + np). \tag{15}$$

These rules associate to any pair of such expressions an expression of the same form. It can be checked that

if $m - n \equiv m' - n'$ and $p - q \equiv p' - q'$, then
$$(m - n) + (p - q) \equiv (m' - n') + (p' - q') \text{ and}$$
$$(m - n) \cdot (p - q) \equiv (m' - n') \cdot (p' - q'). \tag{16}$$

Therefore, the operations (14) and (15) are compatible with the "identifications" and may be considered as operations in $\mathbb{Z}$.

The order relation may be extended to $\mathbb{Z}$, using the definition

$$m - n < p - q \quad \text{if } m + q < n + p. \tag{17}$$

This order relation is also compatible with the identification:

if $m - n \equiv m' - n'$ and $p - q \equiv p' - q'$, then
$$(m - n) < (p - q) \Rightarrow m' - n' < (p' - q'). \tag{18}$$

If we identify a positive integer m with (any and all of) the expressions $(m + n) - n$, $n \in \mathbb{N}$, then the operations (14) and (15) are compatible with the operations in $\mathbb{N}$, so $\mathbb{N}$ may be considered as a certain *subset* of $\mathbb{Z}$. The properties A1, A2, M1, M2, D, O1, O2 can be proved for $\mathbb{Z}$, using the properties for $\mathbb{N}$ and the definitions. The important point is that $\mathbb{Z}$ has additional properties, also provable, that are not true for $\mathbb{N}$. (At the risk of introducing confusion, we now let z denote an arbitrary element of $\mathbb{Z}$.)

A3: Neutral element for addition. *There is an element 0 with the property that $z + 0 = z$, all $z \in \mathbb{Z}$.*

A4: Additive inverses. *For each $z \in \mathbb{Z}$ there is an element $-z$ with the property that $z + (-z) = 0$.*

The role of the neutral element is played by any of the expressions $m - m$, and the role of the additive inverse of $m - n$ is played by $n - m$, or by any other expression $n' - m'$ with the property that $m + n' = n + m'$.

(Any set A that has an operation of addition that satisfies A1–A4 is called a *commutative group*. If A also has an operation of multiplication and satisfies M1, M4, and D as well, then it is called a *commutative ring*. If one drops the commutativity of multiplication, M2, one has a plain *ring*.)

There is an interplay between the order relation and the algebraic operations in $\mathbb{Z}$, summarized in two properties that can be derived using (some of) A1–A4, M1, M2, D, O1, O2, and (12).

O3: Order and addition. *If $m < n$, then $m + p < n + p$.*

O4: Order and multiplication. *If $m < n$ and $p > 0$, then $mp < np$.*

This type of method can be extended to a construction of the rational numbers as well. Consider expressions of the form m/n, where n is a positive integer and m is any integer; this expression represents the "solution" x of the equation $nx = m$. Again it is necessary to introduce an identification:

$$m/n \equiv m'/n' \quad \text{if } mn' = nm'. \tag{19}$$

The set $\mathbb{Q}$ of rational numbers can be thought of as the set of expressions $m - n$, subject to this identification rule.

Addition and multiplication in $\mathbb{Q}$ are *defined* by

$$\frac{m}{n} + \frac{p}{q} = \frac{mq + pn}{nq} \tag{20}$$

$$\frac{m}{n} \cdot \frac{p}{q} = \frac{mp}{nq}, \tag{21}$$

and the order relation is defined by

$$\frac{m}{n} < \frac{p}{q} \quad \text{if } mq < np. \tag{22}$$

The operations (20) and (21) and the order (22) are compatible with the identification rule (19), so the operations and order may be viewed as being defined in $\mathbb{Q}$.

We may consider $\mathbb{Z}$ as a *subset* of $\mathbb{Q}$ by identifying $m \in \mathbb{Z}$ with the expressions mn/n, where n belongs to $\mathbb{N}$. The operations and order in $\mathbb{Z}$ are consistent with those in $\mathbb{Q}$ under this identification. The set $\mathbb{Q}$ with the operations (20), (21), and the order relation (22) has all the preceding properties A1–A4, M1, M2, D, O1–O4. Again, there are new algebraic properties. Here r denotes an element of $\mathbb{Q}$:

M3: Neutral element for multiplication. *There is an element* $\mathbf{1}$ *such that* $r \cdot \mathbf{1} = r$, *all* r.

M4: Multiplicative inverses. *For each* $r \neq 0$, *there is an element* r^{-1} *such that* $r \cdot r^{-1} = \mathbf{1}$.

In fact, the multiplicative neutral element is represented by any n/n, the rational that is identified with the integer 1. A multiplicative inverse of m/n is n/m if $m > 0$, or $(m/n)^{-1} = (-n)/(-m)$ if $m < 0$.

There is another important property to be noted concerning $\mathbb{Q}$.

O5: The Archimedean property. *If* r *and* s *are positive rationals, then there is a positive integer* N *such that* $Nr > s$.

(If we think of s as the amount of water in a bathtub and r as the capacity of a teaspoon, this says that we can bail the water from the bathtub with the teaspoon in at most N steps. Of course N may be large.) To verify O5, suppose that $r = m/n$ and $s = p/q$. Then $Nr = (Nm)/n$, and, in view of (22), we need to find N so that Nmq is larger than np. Obviously $N = np + 1$ will do.

As we have noted, we need to go from $\mathbb{N}$ to $\mathbb{Z}$ to $\mathbb{Q}$ in order to guarantee that simple algebraic equations like $a + x = b$ and $ax = b$ have solutions. However, $\mathbb{Q}$ is still not rich enough to do more interesting algebra. *In fact, the equation* $r^2 = 2$ *does not have a solution* $r \in \mathbb{Q}$. Suppose that it *did* have a solution $r = p/q$, where p and q are integers and q is positive. We may assume that r is in lowest terms, that is, that p and q have *no common factors*. Then $p^2 = 2q^2$, so p is even. Thus, $p = 2m$ with m an integer. Then $4m^2 = 2q^2$, so $2m^2 = q^2$, so q is also even, and so p and q have the common factor 2, a contradiction.

Now it is possible to find an increasing sequence of rationals

$$1, \quad 1.4, \quad 1.41, \quad 1.414, \quad 1.4142, \quad 1.41421, \quad 1.414213, \quad \ldots$$

whose squares get "arbitrarily close" to 2. (The reader is invited to formulate a more precise form of this statement.) As before, we would like to be able to assert that (a) this sequence has a number x as it limit and (b) $x^2 = 2$.

Note what has happened in this section: In effect, we took the positive integers $\mathbb{Z}$ and their operations as raw material and sketched how to *construct* the remaining integers and the rationals. The construction allows us to *prove* the various algebraic and order properties of $\mathbb{Z}$ and $\mathbb{Q}$ from properties of $\mathbb{N}$. In the next section we sketch a construction of the real numbers from the rationals, in order to fill in the gaps like $\sqrt{2}$ and π.

Exercises

1. Prove the assertion (16).
2. Use the definition (14) and the identification (13) to prove A1 and A2 for the integers $\mathbb{Z}$.
3. Use the definition (15) and the identification (13) to prove M1 and M2 for the integers $\mathbb{Z}$.
4. Verify A3 and A4 for the integers $\mathbb{Z}$.
5. Prove from axioms A1–A4 that 0 is unique: If $z + 0' = z$, then $0' = 0$.
6. Prove from A1–A4 that, given integers m and n, the equation $m + x = n$ has a unique solution $x \in \mathbb{Z}$.
7. Use (12) and the remaining axioms for $\mathbb{Z}$ to prove O3 and O4 for $\mathbb{Z}$.
8. Prove the analogue of assertion (16) for the rationals $\mathbb{Q}$.
9. Use the definition (20) and the identification (19) to prove A1 and A2 for $\mathbb{Q}$.
10. Use the definition (21) and the identification (19) to prove M1 and M2 for $\mathbb{Q}$.
11. Use axioms A, M, and D to prove that $r \cdot 0 = 0$.
12. Prove from axioms A, M, and D that, for any rationals r, s, if $r \neq 0$, then the equation $rx = s$ has a unique rational solution x.
13. Prove that there is no rational r such that $r^2 = 3$.
14. Prove that there is no rational r such that $r^3 = 2$.

1C*. The Line and Cuts

The usual geometric representation of the various number systems above uses a horizontal line. Imagine such a line with one point marked as the origin. Choose a unit of length, and march to the right from the origin in steps of unit length, denoting the corresponding points as $1, 2, 3, \ldots$. Similarly, points obtained by going to the left from the origin in steps of unit length are denoted $-1, -2, -3, \ldots$. This gives us a representation of $\mathbb{Z}$. The order relation $p < q$ has the geometric meaning that p is to the left of q. The distance between p and q is the absolute value $|p - q|$. Points corresponding to the remaining rationals are easily introduced: If we divide the interval with endpoints 3 and 4 into five equal subintervals, the first of these has endpoints 3 and $3 + 1/5 = 16/5$ and so on.

The integers $\mathbb{Z}$ determine a partition of the line into disjoint half-open intervals I_n, where $I_n = [n, n+1)$ consists of all points that lie at or to the right of n but strictly to the left of $n+1$. Given a point x of the line, the *integral part* of x, denoted $[x]$, is the unique integer $[x] = n$ such that x belongs to I_n: The point x is at or to the right of the integer n, and its distance from n is less than 1. It is important to note that this may be sharpened if we proceed to the rationals: *Each point x on the line may be approached as closely as we like by rational points.* In fact, partition the interval with endpoints $[x]$ and $[x]+1$ into 10^k equal subintervals. The point x lies in one of these subintervals and is therefore at distance less than 10^{-k} from one of the endpoints of that subinterval, which are rational numbers.

The preceding is the basis for one of the first *constructions* of the reals, due to Dedekind. We want the reals to account for all points on the line, and we want the line to have no gaps: Any sequence of points moving to the right, but staying to the left of some fixed point, should have a limit. We also want to extend the algebraic operations and the order relation to this full set of points, again so that $<$ means "to the left of."

Our starting point for this process can only be the rationals themselves; they must be the scaffolding on which the real numbers are constructed. In order to see how to proceed, we begin by *imagining that the goal has already been accomplished.* Then, for any point x in the line, we could associate to the point, or real number, x a *set S* of rationals – *all rationals that lie strictly to left of x*. If x and x' are distinct points, then there is a rational r strictly between them. (Choose k so large that $1/10^k$ is smaller than the distance between x and x', and look at the rational points $m/10^k$, $m \in \mathbb{Z}$.) Therefore, the set S that corresponds to x and the set S' that corresponds to x' are different: r belongs to S' but not to S. Notice also that the set S that corresponds to x has the following properties:

(C1) S is not empty and is not all of $\mathbb{Q}$.
(C2) If r is in S, s is in $\mathbb{Q}$, and $s < r$, then s is in S.
(C3) S has no largest element.

We call a subset of the rationals that has these three properties, (C1), (C2), and (C3), a *cut*. (Actually, Dedekind considered both S and the set T consisting of all rationals to the right of x; the pair together partitions the rationals not equal to x into two subsets that correspond to the act of cutting the line at the point x.)

Now, conversely, suppose that S is a cut, a subset of the rationals that has the three properties (C1), (C2), and (C3). Then we expect there to be a unique point x such that S consists precisely of the rationals strictly to the left of x. To see this, construct a sequence of rationals as follows. Conditions (C1) and (C2) imply that there is a largest integer r_0 such that $r_0 \in S$. Then $r_0 + 1$ is not in S. Next, there is an integer p, $0 \le p \le 9$ such that

$$r_1 = r_0 + \frac{p}{10} \in S, \qquad r_0 + \frac{p+1}{10} \notin S.$$

Continuing in this way, we can produce a sequence of rationals r_n such that $r_0 \leq r_1 \leq r_2 \leq \ldots$ and

$$r_n \in S, \qquad r_n + \frac{1}{10^n} \notin S.$$

Because of the no-gap condition, we expect this sequence to have a limit x. (In fact, the r_n's are successive parts of what should be the decimal expansion of x.) A bit of thought shows that a rational should belong to S if and only if it is smaller than some r_n, which is true if and only if is to the left of x.

Our discussion to this point says that *if* we had attained our goal, *then* there would be a 1–1 correspondence between real numbers on one hand and cuts on the other. We now *turn the procedure around*. We take as our *objects* the cuts themselves – the subsets of $\mathbb{Q}$ that satisfy (C1), (C2), and (C3). One can introduce algebraic operations and an order relation among the cuts and demonstrate the properties listed in the previous section. For example, the *sum* of two cuts S and S' is defined to be the set of rationals

$$S + S' = \{s = r + r' : r \in S, \; r' \in S'\}. \tag{23}$$

The rational r can be identified with the cut it induces, which we denote by r^*:

$$r^* = \{s \in \mathbb{Q} : s < r\}. \tag{24}$$

In particular, 0^* turns out to be the neutral element for addition of cuts, and 1^* the neutral element for multiplication of cuts.

The order relation is simple: If S and S' are cuts, then we set $S < S'$ if $S \subset S'$ and $S \neq S'$. (The notation $S' \subset S$ means that S' is a subset of S, but not necessarily a proper subset.)

Defining multiplication of cuts is a bit tricky. (The obvious simple adaptation of the sum rule has a problem: The product of two very negative numbers is very positive.) The usual practice is to start by finding a good definition for $S \cdot S'$ when S and S' are both positive, that is, $0^* < S, 0^* < S'$. [The reader may try to find such a definition and to verify the multiplicative and distributive properties M1–M4, D; see the exercises.]

One can verify that the set of all cuts, with the indicated addition and order relation (and the multiplication to which we have merely alluded), satisfies all the properties listed in the previous section. Of course $\mathbb{Q}$ already had all these properties. The key here is that *the set of all cuts satisfies the no-gap condition*. We do not verify this in detail here, because we have not yet defined what we mean for a sequence to have a limit, but it is easy to specify what the limit is. Suppose that $\{S_n\}$ is a sequence of cuts that is increasing and bounded above:

$$S_1 \subset S_2 \subset S_3 \subset \ldots \subset S_n \subset T, \qquad \text{all } n, \tag{25}$$

for some fixed cut T. Then one can show that the union

$$S = \bigcup_{n=1}^{\infty} S_n = S_1 \cup S_2 \cup S_3 \cup \cdots \tag{26}$$

is a cut and should be considered to be the limit of the cuts S_n.

The same idea leads to the proof of a second version of the no-gap condition, the Least Upper Bound Property, stated in the next chapter. Thus, what we have done in this section is to indicate how one can, starting with $\mathbb{Q}$, construct a collection of objects $\mathbb{R}$ that satisfies all the conditions listed in the next section and that contains (a copy of) $\mathbb{Q}$ itself.

Exercises

1. Suppose that S and S' are two cuts. Prove that either $S \subset S'$ or $S' \subset S$, or $S = S'$.
2. Suppose that S and S' are cuts. Prove that $S + S'$ is a cut, that is, that it is a subset of the rationals that has the three properties (C1), (C2), and (C3).
3–6. Prove some or all of the addition properties A1–A4 for cuts.
7. Check the compatibility of addition for rationals and the corresponding cuts: $(r+s)^* = r^* + s^*$.
8. Define the product of positive cuts. Check that your definition gives a cut, and that $S \cdot 1^* = \alpha$ for every nonnegative cut S.
9. Check the compatibility of multiplication for positive rationals and the corresponding cuts.
10–12. Prove some or all of M1, M2, and D for positive cuts.
13. Define the absolute value $|S|$ of a cut and use it to extend the definition of the product to any two cuts. (Hint: Define the product with 0^* separately.)
14–18. Prove some or all of axioms M and D for arbitrary cuts.
19. Suppose that $S_1, S_2, \ldots$ and T are cuts that satisfy (25). Prove that the set S in (26) is a cut, and that it is the smallest cut such that every S_n is smaller than S.
20. Another approach to constructing the reals is to take all formal decimal expansions. (To remove ambiguities like 1 versus .9999..., we could take *nonterminating* formal decimal expansions.) Discuss the difficulties in defining the algebraic operations. For example, what would be the first term in the decimal expansion corresponding to

$$.997999194\ldots + .002000805\ldots ?$$

Is there some stage (preferably specifiable in advance) at which you would be sure to have enough information to know whether the sum is greater than 1?

1D. Proofs, Generalizations, Abstractions, and Purposes

Why do we want proofs? Consider assertions like (4). This is actually an infinite family of assertions, one for each positive integer n. Any single one of these

1D. Proofs, Generalizations, Abstractions, and Purposes

assertions could be checked by performing the required arithmetic. If one checked the first thousand or so, one might become quite confident of the rest, but that is not sufficient for mathematical *certainty*. (There are statements that are valid up to very large integers but not for all integers. A simple example: "n is not divisible by 10^{100}.") Certainty can be established in at least two ways. One way is by *mathematical induction*: Statement (4) is clearly true if $n = 1$, and by adding $(-2)^{n+1}$ to each side and regrouping the right side one obtains the truth of each subsequent statement from the truth of the one that precedes it. Since the first statement is true, so is the second; since the second statement is true, so is the third; and so on. On the other hand, (4) is a special case of the more general sequence of statements (5). Now (5) can also be proved by mathematical induction. *Another way* to prove (5) for any given positive integer n is to multiply both sides by $1 - r$.

Not only do general statements like (5) need to be proved if we are to rely on them, but, as we have just remarked, there may be more than one way to prove such a statement. In addition to general techniques that work in many cases, like mathematical induction, there are specialized tricks that may give more insight into particular problems. We can deduce (4) as a special case of (5) – but only if (5) has been soundly demonstrated. Because of the cumulative nature of mathematics, we want to be very careful about each step we take.

What makes a proof a proof? A proof is simply an argument that is designed to convince, to leave no doubt. A proof by induction is very convincing if it is carried out carefully – and if the listener or the reader is familiar with the technique and has confidence in it. Such a proof demands some sophistication of both the presenter and the presentee. The second proof of (5) mentioned above is clear and convincing to anyone who is comfortable with algebraic manipulation, and may even suggest how (5) was discovered.

How does one learn to "do" proofs? By observation and practice, practice, practice.

The purpose of this book is to proceed along the path from properties of the number system to the most important results from calculus of one variable, with each step justified, and with enough side excursions to keep the walk interesting. *Definitions* are crucial. They give precise meaning to the terms we use. Many results follow fairly directly from the definitions and a bit of logical thinking. One thing to keep in mind: *The more general the statement of a result*, typically *the simpler its proof must be*. The reason is that the proof cannot take advantage of any of those features of special examples that have been abstracted (i.e., removed) in defining the general concepts.

Proving the equality of real numbers (or of sets) is often best accomplished by proving two inequalities: We may prove that number $a =$ number b or that set

$A = $ set B by proving

$$a \leq b \quad \text{and} \quad b \leq a,$$

or

$$A \subset B \quad \text{and} \quad B \subset A.$$

In some brief excursions, in order to get some more interesting results or examples, we will break the logical development and use things like the integral and the natural logarithm before they have been introduced rigorously. No circularity is involved: These results will not be used to develop the later theory.

Exercise

1. Another way to show equality: Prove that the real numbers a and b are equal if and only if, for each positive real ε, the absolute value $|a - b|$ satisfies $|a - b| < \varepsilon$.

2
The Real and Complex Numbers

The previous chapter was somewhat informal. Starting in this chapter we develop the subject systematically and (usually) in logical order. This does not mean that every step in every chain of reasoning will be written out and referred back to the axioms or to results that have already been established. Such a procedure, though possible, is extremely tedious. The goal, rather, is to include enough results – and enough *examples* of reasoning – so that it may be clear how the gaps might be filled.

2A. The Real Numbers

Our starting point is the real number system $\mathbb{R}$. This is a set that has two algebraic operations, addition and multiplication, and an order relation $<$. Let a, b denote arbitrary elements of $\mathbb{R}$. Addition associates to any pair a, b a real number denoted $a + b$, while multiplication associates a real number denoted $a \cdot b$ or simply ab. That $<$ is a *relation* simply means that certain ordered pairs (a, b) of elements of $\mathbb{R}$ are selected, and for these pairs (only) we write $a < b$. These operations and the order relation satisfy the following axioms, or conditions, in which a, b, c denote arbitrary elements of $\mathbb{R}$.

A1 $(a + b) + c = a + (b + c)$.
A2 $a + b = b + a$.
A3 *There is an element* 0 *such that, for all* a, $a + 0 = a$.
A4 *For each* $a \in \mathbb{R}$ *there is an element* $-a \in \mathbb{R}$ *such that* $a + (-a) = 0$.
M1 $(ab)c = a(bc)$.
M2 $ab = ba$.
M3 *There is an element* $1 \neq 0$ *in* $\mathbb{R}$ *such that, for all* a, $a \cdot 1 = a$.
M4 *For each* a *such that* $a \neq 0$, *there is an element* $a^{-1} \in \mathbb{R}$ *such that* $a \cdot a^{-1} = 1$.
D $(a + b)c = ac + bc$; $a(b + c) = ab + ac$.
O1 *For any* a *and* b, *exactly one of the following is true:* $a < b$, $b < a$, *or* $a = b$.
O2 *If* $a < b$ *and* $b < c$, *then* $a < c$.

O3 *If $a < b$, then $a + c < b + c$.*
O4 *If $a < b$ and $0 < c$, then $ac < bc$.*
O5 *If $0 < a$ and $0 < b$, then there is a positive integer n such that $b < a + a + a + \cdots + a$ (n summands).*

There is one more axiom that is satisfied by $\mathbb{R}$, but to state it we need to define some terms. Note, by the way, that we may refer to the elements of $\mathbb{R}$ as numbers or as *points* (thinking of the representation of $\mathbb{R}$ as a line). Also, we write $a \leq b$ to mean that either $a < b$ or $a = b$, we write $b > a$ if $a < b$, and we write $b \geq a$ if $a \leq b$.

Definition. Suppose that A is a nonempty subset of $\mathbb{R}$. A number $b \in \mathbb{R}$ is said to be an *upper bound* for A if for every a in A we have $a \leq b$. If A has an upper bound, then it is said to be *bounded above*.

The number b is said to be a *least upper bound* for A if it is an upper bound and if $b \leq b'$ for every upper bound b'. There is at most one such number, and it is also called the *supremum* of A and denoted

$$b = \sup A.$$

The definitions of *lower bound, bounded below*, and *greatest lower bound* or *infimum* are defined similarly, with $\leq$ replaced by $\geq$. The infimum c, when it exists, is denoted

$$c = \inf A.$$

The last property in our list of properties of $\mathbb{R}$ is the *Least Upper Bound Property*.

O6 If A is any nonempty subset of $\mathbb{R}$ that is bounded above, then there is a least upper bound for A.

[There is an apparent asymmetry here, but property O6 implies the similar *Greatest Lower Bound Property*, and vice versa. In fact, if A is nonempty and bounded below, then its greatest lower bound is precisely the supremum of the set of its lower bounds, and so on. See Exercise 4.]

All the usual algebraic rules for manipulating real numbers and solving simple equations and inequalities can be deduced from these axioms. Here are some examples.

Proposition 2.1

(a) *Given a and b in $\mathbb{R}$, there is a unique $x \in \mathbb{R}$ such that $a + x = b$.*
(b) *Given a and b in $\mathbb{R}$, if $a \neq 0$, then there is a unique $y \in \mathbb{R}$ such that $ay = b$.*

Proof: If $a + x = b$, then

$$b + (-a) = (a + x) + (-a) = (x + a) + (-a) = x + (a + (-a)) = x + 0 = x;$$

so if there is a solution x, then it is unique: $x = b + (-a)$. On the other hand,

$$a + (b + (-a)) = a + ((-a) + b) = (a + (-a)) + b = 0 + b = b + 0 = b,$$

so $x = b + (-a)$ is a solution. A similar argument shows that if $a \neq 0$, then $y = a^{-1}b$ is the unique solution to $ay = b$. $\square$

Corollary 2.2

(a) *The additive and multiplicative neutral elements 0 and 1 are unique.*
(b) *The additive inverse $-a$ and (if $a \neq 0$) the multiplicative inverse a^{-1} are unique.*
(c) *For any a, $-(-a) = a$; if $a \neq 0$, then $(a^{-1})^{-1} = a$.*
(d) *For any $a \in \mathbb{R}$, $a \cdot 0 = 0$ and $(-1) \cdot a = -a$.*

Proof: According to Proposition 2.1, $a + x = a$ and $ay = a$ (if $a \neq 0$) have *unique* solutions; this proves (a). Part (b) also follows from uniqueness. Part (c) follows from commutativity and uniqueness. Finally,

$$a \cdot 0 + 0 = a \cdot 0 = a \cdot (0 + 0) = a \cdot 0 + a \cdot 0;$$
$$a + (-1) \cdot a = 1 \cdot a + (-1) \cdot a = (1 + (-1)) \cdot a = 0 \cdot a.$$

Uniqueness implies that $a \cdot 0 = 0$, which, in turn, implies that $(-1) \cdot a = -a$. $\square$

With these results as encouragement we streamline things by (usually) writing $b - a$ for $b + (-a)$, (often) writing $1/a$ for a^{-1}, and so on. Next we consider some properties of the order relation.

Proposition 2.3

(a) *For any $a \in \mathbb{R}$, exactly one of the following is true: $0 < a$, $0 < (-a)$, or $a = 0$. Moreover, $0 < a$ if and only if $-a < 0$.*
(b) *For any a and b in $\mathbb{R}$, $a < b$ if and only if $0 < b - a$.*

Proof: This uses O3. If $a < 0$, then $0 = a + (-a) < 0 + (-a) = -a$. This argument implies both statements of part (a). If $a < b$, add $-a$ to both sides. If $0 < b - a$, add a to both sides. $\square$

We say that $a \in \mathbb{R}$ is *positive* if $0 < a$ and *negative* if $a < 0$.

Proposition 2.4. *If a and b are positive, then the sum $a + b$ and the product ab are positive.*

Proof: Using O2 and O3, we see that $0 < a$ implies $b < a + b$, while $0 < b$ and $b < a + b$ imply $0 < a + b$. As for the product, by O4 and Corollary 2.2(d), $0 = a \cdot 0 < a \cdot b$. □

Because of transitivity (O2), it is reasonable to write chains of inequalities like $a < b < c < \cdots$.

Corollary 2.5. $0 < 1 < 1 + 1 < 1 + 1 + 1 < \cdots$.

Proof: By the preceding results, either 1 or -1 is positive, and if -1 were positive, then so would be $(-1)(-1) = -(-1) = 1$. Thus 1 must be positive and the remaining inequalities follow from successive applications of O3. □

The following is a particularly important fact about the ordering, which will be used very frequently.

Proposition 2.6. *If $0 < a < b$, then $0 < 1/b < 1/a$.*

Proof: If $1/b$ were negative, we could multiply the inequality $1/b < 0$ by b and conclude that $1 < 0$. Similarly, $1/a$ and $1/ab$ are positive. The second statement follows, since $1/a - 1/b = (b - a)/ab$ is the product of positive numbers. □

Remarks. 1. Axioms A1–A4, M1–M4, and D are the axioms for a *field*. Axiom A3 requires that a field have at least one element, 0, and M3 requires that it have at least one additional element, 1. In fact, there is a field that has exactly two elements, denoted 0 and 1. Addition and multiplication are determined in part by the axioms ($0 + 1 = 1$ and so on) and completed by $1 + 1 = 0$ and $0 \cdot 1 = 0 = 1 \cdot 0$.

2. Axioms A1–A4, M1–M4, D, and O1–O4 are the axioms for an *ordered field*. An ordered field must have infinitely many elements. See Corollary 2.5.

3. So far we have taken the reals as simply an abstract set, with two operations and an order, that satisfies the preceding axioms. Thus, a priori it has no relation to $\mathbb{Q}$ or even to $\mathbb{N}$ and $\mathbb{Z}$. Here is how to remedy that situation. Suppose that n is a positive integer and a belongs to $\mathbb{R}$. Let na or $n \cdot a$ denote $a + a + \cdots + a$, where there are n summands. If n is a negative integer, we let $n \cdot a$ denote $-(-n)a$. (We are letting 1 denote either the multiplicative neutral element of $\mathbb{Q}$ or that of $\mathbb{R}$, but context should make clear which is meant.) We can assign to a rational number

$r = m/n$ the real number $(m \cdot 1)(n \cdot 1)^{-1}$, which we denote by $\tilde{r}$. With some labor one can prove the following, for any rationals r and s.

(a) $\tilde{r} = \tilde{s}$ if and only if $r = s$.
(b) $\widetilde{(r+s)} = \tilde{r} + \tilde{s}$ and $\widetilde{(rs)} = \tilde{r}\tilde{s}$.
(c) $r < s$ if and only if $\tilde{r} < \tilde{s}$.

In other words, the subset $\tilde{\mathbb{Q}} = \{\tilde{r} : r \in \mathbb{Q}\}$ of $\mathbb{R}$, using the operations and order from $\mathbb{R}$, is an exact copy of $\mathbb{Q}$. From now on we *identify* $\mathbb{Q}$ with this copy and consider it to be a subset of $\mathbb{R}$.

4. *The preceding axioms characterize* $\mathbb{R}$. This means that if $\mathbb{R}'$ were another set having two operations and an order that satisfied all the preceding axioms, then there would be a 1–1 correspondence between elements $\mathbb{R}$ and elements of $\mathbb{R}'$ that takes sums to sums, products to products, and preserves the order relation. In other words, $\mathbb{R}$ and $\mathbb{R}'$ can be regarded as identical. [Here is how this result is proved. First, as before, consider $\mathbb{Q}$ as a subset of $\mathbb{R}$. Any given $a \in \mathbb{R}$ can be obtained as a supremum of a set of rationals as follows:

$$a = \sup\{r \in \mathbb{Q} : r < a\}. \tag{1}$$

Now, if $\mathbb{R}'$ is a second such set, we may first identify a subset of it with the rationals and then use (1) to see how to associate to any $a \in \mathbb{R}$ a corresponding element $a' \in \mathbb{R}'$.]

At the end of Section 1B we showed that there is no rational solution of the equation $x^2 = 2$. The situation is different in $\mathbb{R}$.

Theorem 2.7: Existence of *n*-th roots. *Suppose that b is a positive real number and n is a positive integer. There is a unique positive real number a such that $a^n = b$.*

Proof: For any real numbers x and y,

$$y^n - x^n = (y - x)(y^{n-1} + y^{n-2}x + y^{n-3}x^2 + \ldots + yx^{n-2} + x^{n-1}). \tag{2}$$

[Notice that this formula implies the formulas at the beginning of Section 1A!] If $0 < x < y$, then it follows from (2) that $y^n - x^n$ is the product of positive factors and therefore is positive. Therefore, there can be at most one positive solution to $a^n = b$. To show that there is a solution, we note that the set A below is bounded above (prove!) and take advantage of the Least Upper Bound Property. Set

$$a = \sup A, \quad \text{where } A = \{x \in \mathbb{R} : x^n < b\}. \tag{3}$$

We show that $a^n = b$ by showing that a^n cannot be smaller or larger than b.

First, suppose that x is positive and $x^n < b$. We shall choose $y > x$ such that $y^n < b$. This will show that x is not an upper bound for the set A, so $x \neq a$ and therefore $a^n \geq b$. For this purpose, we may assume to begin with that $y \leq x + 1$. It follows from this and from (2) that

$$y^n = x^n + (y^n - x^n) \leq x^n + (y - x)\left[n(x+1)^{n-1}\right], \qquad (4)$$

since there are n terms in the summation on the right-hand side of (2). Therefore, if $x < y \leq x + 1$ and also

$$y - x < \frac{b - x^n}{n(x+1)^{n-1}}, \qquad (5)$$

it follows that $y^n < b$. To accomplish this we can take y to be the smaller of the numbers $x + 1$ and $x + \frac{1}{2}c$, where c is the right-hand side of (5). Thus $a^n \geq b$.

Finally, suppose that y is positive and $y^n > b$. If we show that there is a positive x such that $x < y$ and $b < x^n$, it follows that x is an upper bound for A. But then y is not the least upper bound, so $y \neq a$ and consequently $a^n \leq b$. The proof that such a number x can be chosen is similar to the previous part of the proof and is left as an exercise. $\square$

Remark. There are other (and, in some respects, better) ways to prove Theorem 2.7. However, any proof relies ultimately on the Least Upper Bound Property (or some equivalent property) of the reals. The advantage of this proof is that it makes that reliance explicit and does not depend on introducing notions like continuity and connectedness.

Exercises

1. Show that, for any positive reals x and y, there is a positive integer n such that $x/n < y$.
2. Show that, for any reals x and y with $x < y$, there are a rational r and an irrational t such that $x < r < y$ and $x < t < y$.
3. Show that the Archimedean axiom O5 follows from the Least Upper Bound Property O6, together with the other axioms for the reals.
4. (a) Suppose that A and B are nonempty subsets of $\mathbb{R}$. Define subsets $-A = \{-x : x \in A\}$ and $A + B = \{x + y : x \in A \text{ and } y \in B\}$. Show that if A and B are bounded above, then $\inf(-A) = -\sup(A)$ and $\sup(A + B) = \sup(A) + \sup(B)$.
   (b) Use part (a) to prove the Greatest Lower Bound Property: Any nonempty subset of $\mathbb{R}$ that is bounded below has a greatest lower bound.
5. *The Nested Interval Property*: Suppose that $I_1, I_2, I_3, \ldots$ is a sequence of bounded closed intervals of reals, $I_n = [a_n, b_n]$, where $a_n \leq b_n$. Suppose that $I_1 \supset I_2 \supset \ldots \supset I_n \ldots$, and suppose that the lengths $|I_n| = |b_n - a_n|$ have limit zero. (This means that, for any $\varepsilon > 0$, there is an integer N such that $|I_n| < \varepsilon$ if $n \geq N$.) Show that there is exactly one real number x that belongs to all the intervals.

6. Assume the axioms for the reals *except* for the Least Upper Bound Property O6; assume instead the Nested Interval Property, formulated in Exercise 5. Prove the Least Upper Bound Property as a consequence.
7. (a) Suppose that $x > 0$ and $x^2 < 2$. Prove that there is a real $y > x$ such that $y^2 < 2$. Show that y may be chosen to be rational.
   (b) Suppose that $x > 0$ and $x^2 > 2$. Prove that there is a real $0 < y < x$ such that $y^2 > 2$. Show that y may be chosen to be rational.
8. Use the preceding exercise to show that if $a = \sup(A)$, where $A = \{r \in \mathbb{Q} : r > 0 \text{ and } r^2 < 2\}$, then $a^2 = 2$. This demonstrates that $\mathbb{Q}$ does not have the Least Upper Bound Property.
9. Prove that $n \in \mathbb{Z}_+$ implies $2^n > n$; prove that for any integer $n \geq 4$, $2^n \geq n^2$.
10. Prove that for any positive h and any integer $n \geq 0$, $(1 + h)^n \geq 1 + nh$.
11. Prove that for any positive h and any integer $n \geq 0$,
$$(1 + h)^n \geq 1 + nh + \frac{n(n-1)}{2} h^2.$$
12. Suppose that a is positive and $n \geq 2$ is an integer. Suppose that $a^n = n$. Prove that $1 < a < 1 + \sqrt{2/(n-1)}$.
13. Let $N = 10^8$. Compute $N^{1/N}$ to three decimal places.
14. True or false? For every $\varepsilon > 0$ there are positive integers m and n such that the inequality $|\sqrt{n} - \sqrt{m} - \pi| < \varepsilon$ is true.
15. Show that there are no rationals r and s such that $r^2 = 8$ or $s^3 = 6$.
16. Suppose that $x > 0$ and $x^3 + x = 4$. Prove that x is irrational.
17. Show that there is at most one real x such that $x^5 + x^3 + 3 = 0$, and it cannot be rational.
18. There is an investment strategy called "dollar cost averaging" based on the claim that investing a fixed amount of money in a stock at each of n times results in an average price per share that is less than the mean of the prices at the various times. Discuss.

2B*. Decimal and Other Expansions; Countability

Suppose that x is a real number. We may deduce from the Archimedean property O5 and the other properties of the reals that there is a unique integer m such that $m < x \leq m + 1$. Geometrically, this simply says that the half-open intervals $(m, m+1]$ partition the line. (We use here the reasonably standard interval notation:

$$(a, b] = \{x \in \mathbb{R} : a < x \leq b\}, \qquad [a, b] = \{x \in \mathbb{R} : a \leq x \leq b\}, \qquad (6)$$

and so on.) Partitioning the interval $(m, m+1]$ into ten half-open subintervals of length $1/10$, or simply looking at the algebra, we see that there is a unique integer a_1, $0 \leq a_1 \leq 9$, such that $m + a_1/10 < x \leq m + a_1/10 + 1/10$. Continuing, we find a unique sequence of integers $m, a_1, a_2, \ldots$ that is characterized by the properties

$$m + \frac{a_1}{10} + \frac{a_2}{10^2} + \cdots + \frac{a_k}{10^k} < x \leq m + \frac{a_1}{10} + \frac{a_2}{10^2} + \cdots$$
$$+ \frac{a_k}{10^k} + \frac{1}{10^k}, \quad 0 \leq a_j \leq 9. \tag{7}$$

If we also write the integer part m in its decimal form, the result is the *decimal expansion* of x.

For convenience, suppose that $m = 0$, that is, $0 < x \leq 1$. We write, formally,

$$x = .a_1 a_2 \ldots a_k \ldots = \frac{a_1}{10} + \frac{a_2}{10^2} + \cdots + \frac{a_k}{10^k} + \cdots. \tag{8}$$

(Notice that our procedure gives $.999\ldots$ as the decimal expansion of 1, $.4999\ldots$ as the decimal expansion of $1/2$, and so on. In fact, this construction guarantees that there are infinitely many a_k's that are not zero.)

One justification for writing (8) is the following: The sequence of rationals

$$r_1 = \frac{a_1}{10}, \quad r_2 = \frac{a_1}{10} + \frac{a_2}{10^2}, \quad r_3 = \frac{a_1}{10} + \frac{a_2}{10^2} + \frac{a_3}{10^3}, \quad \ldots \tag{9}$$

satisfies

$$r_1 \leq r_2 \leq r_3 \ldots \leq 1; \quad x = \sup\{r_1, r_2, r_3, \ldots\}. \tag{10}$$

Conversely, suppose that $\{a_1, a_2, a_3, \ldots\}$ is a sequence of integers such that $0 \leq a_k \leq 9$ and infinitely many of the a_k's are not zero. Define rationals r_k by (9). These satisfy the first part of (10), and we may use the second part of (10) to *define* a number x. It is not difficult to show that this number x has the expansion (8). In other words *there is a 1–1 correspondence between the real numbers in the interval $(0, 1]$ and the decimal expansions $.a_1 a_2 \ldots$ that have infinitely many nonzero terms.* Geometrically, we can look on the decimal expansion as specifying a sequence of *intervals*: The first term identifies that one of the ten equal subintervals of $(0, 1]$ that contains x, the second indentifies that one of the ten equal subintervals of the first that contains x, and so on. Together these successive intervals locate x to any degree of accuracy.

There is nothing special about the number 10 in this discussion, other than the fact that it is an integer larger than 1. If we split successive intervals into two or three equal subintervals and continue as above, we obtain the *binary expansion*

$$x = \frac{b_1}{2} + \frac{b_2}{2^2} + \frac{b_3}{2^3} + \cdots; \quad \text{each } b_k = 0 \text{ or } 1, \tag{11}$$

or the *ternary expansion*

$$x = \frac{c_1}{3} + \frac{c_2}{3^2} + \frac{c_3}{3^3} + \cdots; \quad \text{each } c_k = 0, 1, \text{ or } 2. \tag{12}$$

We have mentioned several times the concept of a 1–1 *correspondence*. This concept allows one to make precise the idea that two sets have the same number of elements. In fact, one can view the usual process of counting as the process of establishing a 1–1 correspondence between the objects being counted and one of the standard sets

$$\emptyset, \ \{1\}, \ \{1, 2\}, \ \{1, 2, 3\}, \ \{1, 2, 3, 4\}, \ \ldots,$$

where $\emptyset$ denotes the *empty set*, the set with no elements. One can extend the idea to infinite sets. In particular, an infinite set is said to be *countable* if it can be put into 1–1 correspondence with the set $\mathbb{N}$ of positive integers. Otherwise, it is said to be *uncountable*.

This notion has some surprising features. The relation $n \leftrightarrow n+1$ establishes a 1–1 correspondence between $\mathbb{N}$ and the "smaller" set consisting of integers ≥ 2, while the relation $n \leftrightarrow 2n$ establishes one with the "much smaller" set of even integers. In the other direction, listing

$$0, \ 1, \ -1, \ 2, \ -2, \ 3, \ -3, \ 4, \ -4, \ldots$$

shows how to establish a 1–1 correspondence between $\mathbb{N}$ and the "much larger" set of all integers; in other words, $\mathbb{Z}$ is countable. In fact $\mathbb{Q}$ *is countable*. It is not difficult to see how to write a list; another approach is to note that the correspondence $2^{m-1}(2n-1) \leftrightarrow m/n$ establishes that the positive rationals are countable.

Not all infinite sets are countable, however. Georg Cantor, who founded set theory, proved the following.

Theorem 2.8. *The set $\mathbb{R}$ of real numbers is uncountable.*

Proof: If $\mathbb{R}$ were countable, then the interval $(0, 1]$ would be countable. If so, then one could make a list of all the corresponding decimal expansions

$$\begin{cases} .a_1 a_2 a_3 a_4 \ldots \\ .b_1 b_2 b_3 b_4 \ldots \\ .c_1 c_2 c_3 c_4 \ldots \\ \quad \ldots \end{cases}$$

However, we can find a number not on the list by constructing its decimal expansion $.x_1 x_2 x_3 x_4 \ldots$ according to the simple rules $x_1 \neq a_1$, $x_2 \neq b_2$, $x_3 \neq c_3$, and so on. This expansion differs in its k-th place from the k-th expansion on the list for every k. (We also choose so that each of these expansions has infinitely many nonzero terms, as does the new one being constructed.) It follows that x is not on the list. □

Exercises

1. Show that any rational in the interval $(0, 1]$ can be expressed as a finite sum $r = 1/q_1 + 1/q_2 + \ldots + 1/q_N$, where the q_j are integers and $q_1 < q_2 < \ldots < q_N$.
2. Discuss the probability that a real number chosen at random from the interval $[0, 1]$ has no 7's in its decimal expansion.
3. Give a proof of Theorem 2.8 using binary expansions. Be careful about the possibility of two binary expansions representing the same real number.

2C*. Algebraic and Transcendental Numbers

We may think of the rational numbers as the subset of $\mathbb{R}$ whose elements are all solutions of equations $nx - m = 0$, where n is a positive integer and m is any integer. More generally, a real number is said to be *algebraic* if it is a solution of a polynomial equation with integer coefficients:

$$a_k x^k + a_{k-1} x^{k-1} + \cdots + a_1 x + a_0 = 0, \qquad a_k \in \mathbb{N}, \quad a_{k-1}, \ldots, a_0 \in \mathbb{Z}. \qquad (13)$$

If x is a solution of (13) but not a solution of any such equation having degree less than k, then x is said to be *algebraic of degree k*. Thus the rationals are precisely the real numbers that are algebraic of degree 1.

So far the only real number that we have shown to be irrational is $\sqrt{2}$. However, $\sqrt{2}$ is a solution of $x^2 - 2 = 0$, so it is algebraic. A real number that is *not* algebraic is said to be *transcendental*. It is not obvious that there are any transcendental numbers. Late in the nineteenth century, the number e was shown to be transcendental by Hermite and π was shown to be transcendental by Lindemann. We give here two proofs that there are transcendental numbers. The first is a counting argument: There are not enough algebraic numbers to account for all real numbers.

Theorem 2.9. *The set of algebraic numbers is countable.*

Proof: Define the *weight* of the polynomial in (13) to be the integer $k + a_k + |a_{k-1}| + \cdots + |a_0| - 1$. Thus the minimum possible weight is 1, the weight of the polynomial x. The polynomials of weight 2 are x^2, $x + 1$, $x - 1$, and $2x$, and those of weight 3 are $x^2 \pm 1$, $x \pm 2$, $2x \pm 1$, and $2x^2$. There are finitely many possible polynomials (13) of any given weight w. Each of these has degree less than w and therefore has at most $w - 1$ roots. This allows us to list the algebraic numbers in a systematic way: For weights 1, 2, 3 we get 0, ± 1, ± 2, $\pm 1/2$. Going to weight 4 we obtain a few irrationals; for example, the polynomials $x^2 - 2$ and $x^2 + x + 1$ have real roots $\pm\sqrt{2}$ and $(-1 \pm \sqrt{5})/2$. □

Theorem 2.9 tells us that in some sense "most" reals are transcendental, but it does not exhibit any transcendental number. The first examples were given by Liouville, based on his theorem about approximation by rationals.

Theorem 2.10. *Suppose that x is an algebraic real number of of degree $k \geq 2$. Then there is a constant $K > 0$ such that, for every rational r, if $r = p/q$ in lowest terms, then*

$$|x - r| \geq \frac{K}{q^k}. \tag{14}$$

Proof: The proof can be made purely algebraic, but it is quicker to appeal to calculus – specifically, the Mean Value Theorem. Notice that (14) is true with any $K \leq 1$ if $|x - r| \geq 1$, so we may assume that $|x - r| \leq 1$.

Let P denote the polynomial in (13) and suppose that $P(x) = 0$. There is a constant L such that the derivative P' satisfies $|P'(s)| \leq L$ if $|s - x| \leq 1$. Now suppose that $|x - r| \leq 1$. The Mean Value Theorem says that

$$P(r) = P(r) - P(x) = (r - x)P'(s), \tag{15}$$

where s is some number between r and x. It follows that

$$|r - x| \geq \frac{|P(r)|}{L}. \tag{16}$$

However, if $r = p/q$, then $q^k P(r)$ is a (nonzero) integer, so $|P(r)| \geq 1/q^k$. Therefore (14) holds, with K taken to be the smaller of 1 and $1/L$. □

Remark. The preceding theorem is also valid for $k = 1$ if we add the conditions that $r = p/q$ is in lowest terms and that q is sufficiently large; this is necessary to rule out $r = x$.

Corollary 2.11. *The number with decimal expansion*

$$.110001000000000000000100000000000000000\ldots \tag{17}$$

(where the k-th one in the expansion occurs in the $k!$ place) is transcendental.

Proof: Let x be the number with decimal expansion (17). We leave it to the reader to check that for any positive integer k and any constant K the sequence of rationals

$$r_1 = \frac{1}{10}, \quad r_2 = \frac{1}{10} + \frac{1}{10^2}, \quad r_3 = \frac{1}{10} + \frac{1}{10^2} + \frac{1}{10^6},$$
$$r_4 = \frac{1}{10} + \frac{1}{10^2} + \frac{1}{10^6} + \frac{1}{10^{24}}, \quad \ldots$$

eventually violates (14). Therefore x is not algebraic of any degree. □

Remark. The algebraic numbers form a field: the sum, product, difference, and quotient of algebraic numbers is algebraic; see the exercises.

Exercises

1. Suppose that x is algebraic of degree n. Prove that $-x$ and (if $x \neq 0$) $1/x$ are also algebraic of degree n.
2. Suppose that x is a real number. Prove that the following are equivalent:
    (i) x is algebraic of degree $\leq n$;
    (ii) $1, x, x^2, \ldots, x^n, x^{n+1}$ are linearly dependent over $\mathbb{Q}$;
    (iii) every nonnegative integer power of x is a linear combination of $1, x, x^2, \ldots, x^n$ with rational coefficients.
    (iv) there are rational numbers $y_1, y_2, \ldots, y_n$ such that every nonnegative integral power of x is a linear combination of the y_k with rational coefficients.
3. Use the preceding exercise to show that $2^{1/3} + 3^{1/2}$ is algebraic of degree at most 6. In fact, if x and y are real numbers that are algebraic of degrees m and n, respectively, then $x + y$ and xy are algebraic of degree $\leq mn$.

2D. The Complex Numbers

To start, we take as the complex numbers the set $\mathbb{C}$ of all expressions $a + ib$, where a and b are real numbers and i is simply a place marker; we could equally well write this as an ordered pair (a, b). In the set $\mathbb{C}$ we *define* operations

$$(a + ib) + (c + id) = (a + c) + i(b + d);$$
$$(a + ib)(c + id) = (ac - bd) + i(ad + bc). \tag{18}$$

Proposition 2.12. $\mathbb{C}$ *satisfies the field axioms A1–A4, M1–M4, and D.*

Proof: Most of this is routine checking. The neutral elements for addition and multiplication respectively are $0 + i0$ and $1 + i0$. If $a + ib \neq 0 + i0$, which is the same as saying $a^2 + b^2 \neq 0$, then the multiplicative inverse is

$$(a + ib)^{-1} = \frac{a}{a^2 + b^2} + i \frac{-b}{a^2 + b^2}. \quad \square \tag{19}$$

2D. The Complex Numbers

The correspondence that assigns to $a \in \mathbb{R}$ the complex number $a + i0$ takes sums to sums, products to products, and the neutral elements for addition and multiplication in $\mathbb{R}$ to the neutral elements of $\mathbb{C}$. Therefore we may *identify* a with $a + i0$ and consider $\mathbb{R}$ as a subset of $\mathbb{C}$. Using the definition (18), one can check easily that $(0 + i1)^2 = -1 + i0$, the complex number identified with the real number -1.

From now on we will make no distinction between a and $a + i0$, and we will also write ib for $0 + ib$ and $\pm i$ for $(\pm 1) \cdot i$. Thus $i^2 = -1$. Moreover, we may denote complex numbers in general by single letters $z, w, \ldots$ or even $a, b, \ldots$. Also, we may denote z^{-1} by $1/z$.

Definitions. Suppose that x and y are real and let $z = x + iy$. Then x and y are called the *real part* and *imaginary part* of z:

$$x = \operatorname{Re} z \qquad y = \operatorname{Im} z.$$

The complex number $\bar{z} = x - iy$ is called the *complex conjugate* of z. The real number $|z| = \sqrt{x^2 + y^2}$ is called the *modulus* of z.

There is a potential conflict of notation here, since a real number x with absolute value $|x|$ can also be considered as a complex number $x = x + i0$ with modulus $|x + i0|$, but in fact these are the *same*, namely, $\sqrt{x^2}$.

Various useful algebraic facts are summed up in the following.

Proposition 2.13. *For any complex numbers z and w,*

$$z = \operatorname{Re} z + i \operatorname{Im} z; \tag{20}$$

$$\operatorname{Re} z = \tfrac{1}{2}(z + \bar{z}), \qquad \operatorname{Im} z = \tfrac{1}{2i}(z - \bar{z}); \tag{21}$$

$$\overline{z + w} = \bar{z} + \bar{w}, \qquad \overline{zw} = \bar{z}\bar{w}; \tag{22}$$

$$|z|^2 = z\bar{z}; \qquad |z| = |\bar{z}|; \tag{23}$$

$$|zw| = |z|\,|w|; \qquad \text{if } w \neq 0, \quad \text{then } |z/w| = |z|/|w|; \tag{24}$$

$$\text{if } z \neq 0, \quad \text{then } z^{-1} = \bar{z}/|z|^2. \tag{25}$$

Proof: Each identity in (20)–(23) is either an easy consequence of the definitions or is easily checked by calculation. The first identity (24) is a consequence of the first identity in (23) and the second identity in (24) follows from the first. The identity (25) follows from the first identity (23). □

We may identify the complex number $z = x + iy$ with the point (x, y) in the (coordinatized) plane $\mathbb{R}^2$. Then the geometric interpretation of complex conjugation is a reflection about the horizontal coordinate axis. The geometric interpretation of

$|z|$ is that it is the Euclidean *distance* from the point z to the origin (Pythagorean Theorem). More generally, $|z - w|$ *is the Euclidean distance between the points z and w*.

The following inequalities are elementary but very important. The third, (28), is known as the *triangle inequality* for the modulus. With the interpretation of the modulus as the distance it may seem geometrically evident, but it is important to prove it.

Proposition 2.14. *For any complex numbers z and w,*

$$|\operatorname{Re} z| \leq |z|, \qquad |\operatorname{Im} z| \leq |z|; \tag{26}$$

$$|z| \leq |\operatorname{Re} z| + |\operatorname{Im} z|; \tag{27}$$

$$|z + w| \leq |z| + |w|. \tag{28}$$

Proof: Let $x = \operatorname{Re} z$, $y = \operatorname{Im} z$. Then $|z| = \sqrt{x^2 + y^2}$ is clearly larger than or equal to both $|x|$ and $|y|$. Conversely, (27) follows from squaring both sides. To prove (28) we square and then use Proposition 2.13 together with (26) to obtain

$$\begin{aligned}|z+w|^2 &= (z+w)(\bar{z}+\bar{w}) = z\bar{z} + (z\bar{w} + w\bar{z}) + w\bar{w} \\ &= |z|^2 + 2\operatorname{Re}(z\bar{w}) + |w|^2 \leq |z|^2 + 2|z\bar{w}| + |w|^2 \\ &= |z|^2 + 2|z||w| + |w|^2 = (|z| + |w|)^2. \quad \square\end{aligned}$$

Remarks. We have not introduced an *order* into $\mathbb{C}$. It is not possible to do so in such a way that $\mathbb{C}$ becomes an ordered field, that is, so that O1–O4 are satisfied (see Exercise 2). *An inequality $z < w$ or $z \leq w$ has no meaning unless w and z are both real*. Note that the inequalities in the preceding proposition all involve real numbers!

From now on we shall frequently use the following convention: Whenever a complex number is written in the form $z = x + iy$, then it is understood that x and y are real, the real and imaginary parts of z.

Exercises

1. Prove that an ordered field cannot be finite.
2. Prove that there is no way to choose an ordering in $\mathbb{C}$ such that it becomes an ordered field.
3. Suppose that $z = x + iy$ with x and y real. Show that $\max\{|x|, |y|\} \leq |z| \leq |x| + |y|$.
4. (a)–(d) Prove Proposition 2.13 in detail.
5. Suppose that z is a nonzero complex number. Show that there is a unique pair $r \in \mathbb{R}$, $w \in \mathbb{C}$ such that $r > 0$, $|w| = 1$, and $z = rw$. This is called the *polar decomposition* of z.

6. Let $S = \{z \in \mathbb{C} : |\text{Im } z| \leq \text{Re } z\}$ and $T = \{z^2 : z \in S\}$. Describe these sets in terms of the modulus and argument of their elements.

7. For complex z, show that $|z - i| = |z + i|$ if and only if z is real; give a geometric interpretation of this result.

8. Suppose that a, b, and c are complex numbers such that $|a| = |b| = |c| = 1$ and $a + b + c = 0$. Prove that $|a - b| = |b - c| = |c - a|$ and discuss the geometric meaning.

9. Let $f(z) = (z - i)/(z + i)$ for complex $z \neq -i$. Show that, for each complex number $w \neq 1$, there is a unique z such that $f(z) = w$. Show that the image of the real line $f(\mathbb{R})$ is the unit circle $\{z : |z| = 1\}$. Find the image of the upper half-plane $\{z : \text{Im } z > 0\}$.

10. (a) Suppose that A and C are real and B is complex. Prove that the set of complex z's such that $Az\bar{z} + \bar{B}z + B\bar{z} + C = 0$ is either a circle, a straight line, or the empty set.
    (b) Prove conversely that any circle or straight line in the complex plane can be described by an equation of this form.

11. Suppose that a, b, c, and d are complex numbers such that $ad - bc \neq 0$. Define a function f by $f(z) = (az + b)/(cz + d)$ whenever $cz + d \neq 0$. Prove that if S is a circle or straight line in the complex plane, then the image $f(S) = \{f(z) : z \in S\}$ is a circle or straight line.

    The following set of exercises gives a nonrigorous preview of material that will be developed rigorously in Chapter 9. Earlier we have used "show" as a synonym for "prove," but in these exercises it has a looser meaning.

12. Use the power series expansions of $\sin t$ and $\cos t$ to justify *defining* $e^{it} = \cos t + i \sin t$ for real t.

13. Show that $|e^{it}| = 1$ and $e^{it+is} = e^{it}e^{is}$.

14. Use Exercises 3, 12, and 13 to show that each complex $z \neq 0$ can be written as re^{it}, where $r > 0, t \in \mathbb{R}$. Relate this to polar coordinates in the plane.

15. Show that for any $n \in \mathbb{N}$ there are exactly n complex solutions to the equation $z^n = 1$.

16. Show that for any complex $w \neq 0$ and any $n \in \mathbb{N}$ there are exactly n complex solutions to the equation $z^n = w$.

3
Real and Complex Sequences

The goal of this chapter is to establish the basic definitions and results concerning the convergence of real and complex sequences. These include monotone sequences, upper and lower limits of real sequences, the Cauchy criterion, and subsequences.

3A. Boundedness and Convergence

A complex *sequence* is a collection of complex numbers $a_1, a_2, a_3, \ldots$ indexed by the integers $1, 2, 3, \ldots$. We denote such a sequence by $\{a_n\}_{n=1}^\infty$ or $\{a_n\}_1^\infty$ or simply by $\{a_n\}$. We also consider sequences $\{a_n\}_{n=0}^\infty$ indexed by the nonnegative integers $0, 1, 2, 3, \ldots$. Such a sequence is said to be *real* if each of its terms a_n is real.

Definitions. A complex sequence $\{a_n\}$ is said to be *bounded* if there is a real $K \geq 0$ such that $|a_n| \leq K$, all n. A real sequence $\{a_n\}$ is said to be *bounded above* if there is a real M such that $a_n \leq M$, all n; it is said to be *bounded below* if there is a real L such that $L \leq a_n$, all n.

The first of the following two sequences is bounded below but not bounded, and the second is bounded:

$$1, 2, 3, 4, \ldots; \qquad 1, 0, -1, 1, 0, -1, 1, 0, -1, \ldots. \tag{1}$$

It is fairly obvious that a real sequence is bounded if and only if it is bounded both above and below. Moreover, a complex sequence $\{a_n\}$ is bounded if and only if the two real sequences $\{\operatorname{Re} a_n\}$ and $\{\operatorname{Im} a_n\}$ are bounded.

Definition. The complex sequence $\{a_n\}$ is said to *converge* or to be a *convergent sequence* if there is a complex number a with the property that for *each $\varepsilon > 0$ there is an integer N such that $n \geq N$ implies $|a_n - a| < \varepsilon$.* [*This is one of the most important definitions in the subject.*] If so, a is said to be the *limit* of the sequence.

3A. Boundedness and Convergence

Remarks. 1. Neither of the sequences (1) has a limit, according to this definition. Later we will modify the definition and give a sense to saying that the first sequence has limit $+\infty$.

2. It can be tempting to think of the preceding definition as saying that the terms of the sequence get *progressively closer* to the limit, but this is not necessarily the case. Let $a_n = 1/n$ if n is an odd positive integer and $a_n = 0$ if n is even, so that the sequence is

$$1, \ 0, \ 1/3, \ 0, \ 1/5, \ 0, \ 1/7, \ 0, \ldots. \tag{2}$$

Then the limit is 0, but each even term is closer to the limit than is *any* odd term.

3. We have been speaking of "the limit." In fact, suppose that both a and a' are limits for the sequence $\{a_n\}$. Given any $\varepsilon > 0$, choose N and N' so that $|a_n - a| < \varepsilon$ if $n \geq N$ and $|a_n - a'| < \varepsilon$ if $n \geq N'$. Choose $n \geq N + N'$. Then

$$|a - a'| = |(a - a_n) + (a_n - a')| \leq |a - a_n| + |a_n - a'| < \varepsilon + \varepsilon = 2\varepsilon.$$

Since this inequality is true for every $\varepsilon > 0$, it follows that $a = a'$. [This method of proving equality is used frequently.] Thus, *if there is a limit, it is unique*.

4. Here are some important examples of convergent sequences. The proofs are left as exercises.

$$\lim_{n \to \infty} 1/n = 0; \tag{3}$$

$$\lim_{n \to \infty} a^n = 0 \quad \text{if } |a| < 1; \tag{4}$$

$$\lim_{n \to \infty} a^{1/n} = 1 \quad \text{if } a > 0; \tag{5}$$

$$\lim_{n \to \infty} n^{1/n} = 1. \tag{6}$$

The question of convergence can be reduced to the question of convergence of real sequences.

Proposition 3.1. *A complex sequence $\{b_n + ic_n\}$ has limit $b + ic$ if and only if the real sequences $\{b_n\}$ and $\{c_n\}$ have limits b and c, respectively.*

Proof: This is a good exercise in using the definitions, together with the inequalities (26) and (27) of Chapter 2. □

Definition. A real sequence $\{a_n\}_{n=1}^{\infty}$ is said to be *nondecreasing* if

$$a_1 \leq a_2 \leq a_3 \leq \ldots \leq a_n, \quad \text{all } n. \tag{7}$$

The sequence is said to be *nonincreasing* if

$$a_1 \geq a_2 \geq a_3 \geq \ldots \geq a_n, \quad \text{all } n. \tag{8}$$

It is said to be *monotone* if it is either nondecreasing or nonincreasing.

Note that a nondecreasing sequence is bounded below (but not necessarily above) and a nonincreasing sequence is bounded above (but not necessarily below).

The following theorem is the starting point for most of the theory of convergence. (Note that it is the "no-gap" condition, as formulated in Chapter 1.)

Theorem 3.2: Convergence of bounded monotone sequences. *Any bounded monotone sequence of reals is convergent.*

Proof: Assume that $\{a_n\}$ is bounded and nondecreasing. Let a be the least upper bound of the set of numbers $\{a_1, a_2, \ldots\}$. If ε is positive, then the number $a - \varepsilon$ is less than a and therefore is not an upper bound for this set. Consequently, there is some index N such that $a_N > a - \varepsilon$. This, the inequalities (7), and the fact that a is an upper bound imply that for any integer $n \geq N$

$$a - \varepsilon < a_N \leq a_n \leq a.$$

Therefore $n \geq N$ implies $|a_n - a| < \varepsilon$, and a is the limit. The nonincreasing case is proved in a similar way. □

We shall see in the next section that a convergent sequence is necessarily bounded, so a monotone sequence is convergent if and only if it is bounded.

Exercises

1. Suppose that $\{x_n\}_1^\infty$ is a sequence of real numbers with limit x, and suppose that $a \leq x_n \leq b$, all n. Prove that $a \leq x \leq b$.
2. Suppose that $z_n = x_n + iy_n$, with x_n and y_n real. Prove that $\lim_{n\to\infty} z_n = x + iy$ with x and y real if and only if $\lim_{n\to\infty} x_n = x$ and $\lim_{n\to\infty} y_n = y$.
3–6. Prove (3) to (6), using Exercises 10–12 of Section 2A.
7. Suppose that $\{z_n\}_1^\infty$ is a complex sequence with limit z_0. Prove directly from the definitions that (a) $\lim_{n\to\infty} z_n^2 = z_0^2$ and (b) $\lim_{n\to\infty} z_n^k = z_0^k$, $k \in \mathbb{N}$.
8. Suppose that $|z| < 1$. Prove that $\lim_{n\to\infty} z^n = 0$
9. Determine: (a) $\lim_{n\to\infty}[\sqrt{n^2 + 2n} - n]$, (b) $\lim_{n\to\infty}(2^n + n)/(3^n - n)$, and (c) $\lim_{n\to\infty}(n!)^{1/n}$.
10. Prove that Theorem 3.2 is equivalent to the Least Upper Bound Property. More precisely, assume all the axioms for the reals from Section 2A except O6, and assume that Theorem 3.2 is valid. Show that O6 is a consequence of these assumptions.

11. Prove that the following limits exist, and evaluate them:

(a) $$\lim_{n \to \infty} \left(\frac{n^2 + 3n + 5}{n^2 + 4n + 7} \right)^n.$$

(b) $$\lim_{n \to \infty} \left\{ \left(1 + \frac{1}{n}\right)\left(1 + \frac{2}{n}\right) \cdots \left(1 + \frac{n}{n}\right) \right\}^{1/n}.$$

12. Suppose that $x_1 = 1$ and $x_{n+1} = \sqrt{2 + x_n}$, $n = 1, 2, \ldots$. Prove that the sequence $\{x_n\}_1^\infty$ converges and find its limit.
13. Suppose that $x_1 = 1$ and $x_{n+1} = 1 + 1/x_n$, $n = 1, 2, \ldots$. Prove that the sequence $\{x_n\}_1^\infty$ converges and find its limit.
14. Let $x_1 = 1$, $x_2 = 2$, and $x_{n+2} = \frac{1}{3}x_n + \frac{2}{3}x_{n+1}$, $n = 1, 2, \ldots$. Prove that the sequence $\{x_n\}_1^\infty$ converges and find its limit.
15. Let positive a and x_1 be given, and let $x_{n+1} = (a + x_n)/(1 + x_n)$, $n = 1, 2, \ldots$. Prove that the sequence $\{x_n\}_1^\infty$ converges and find its limit.
16. Let positive a and x_1 be given, and let $x_{n+1} = \frac{1}{2}(x_n + a/x_n)$, $n = 1, 2, \ldots$. Prove that the sequence $\{x_n\}_1^\infty$ converges and find its limit.

3B. Upper and Lower Limits

Suppose that $\{x_n\}$ is a bounded real sequence. Then all its terms are contained in the closed interval $[a_1, b_1]$, where

$$a_1 = \inf\{x_1, x_2, x_3, \ldots\}; \qquad b_1 = \sup\{x_1, x_2, x_3, \ldots\}.$$

It follows from the definitions that this is the *smallest* closed interval that contains all the terms x_n. Similarly, if we omit the first $m - 1$ terms $x_1, x_2, \ldots, x_{m-1}$, then the remaining terms are contained in a smallest closed interval $[a_m, b_m]$, where

$$a_m = \inf\{x_m, x_{m+1}, x_{m+2}, \ldots\}, \qquad b_m = \sup\{x_m, x_{m+1}, x_{m+2}, \ldots\}. \qquad (9)$$

Consider, as an example, the sequence (2). For it we have $a_n = 0$ for every n, while $b_{2k-1} = b_{2k} = 1/(2k - 1)$, $k = 1, 2, \ldots$. For the second sequence in (1), on the other hand, $a_n = -1$ and $b_n = 1$, all n.

It follows from the definitions that the intervals $[a_n, b_n]$ are nested:

$$a_1 \leq a_2 \leq a_3 \leq \ldots \leq a_n \leq b_n \leq \ldots \leq b_3 \leq b_2 \leq b_1, \qquad \text{all } n. \qquad (10)$$

Thus the nondecreasing sequence $\{a_n\}$ is bounded above by each b_m. According to Theorem 3.2 and its proof, $\{a_n\}$ converges to its least upper bound $a \leq b_m$. Similarly, the nonincreasing sequence $\{b_n\}$ converges and has limit $b \geq a$. These limits are called, respectively, the *lower limit* and *upper limit* of the sequence $\{x_n\}$

and denoted by $\liminf_{n\to\infty} x_n$ and $\limsup_{n\to\infty} x_n$. Summarizing,

$$\liminf_{n\to\infty} x_n = \lim_{n\to\infty} \left[\inf\{x_n, x_{n+1}, \ldots\}\right] \leq \lim_{n\to\infty} \left[\sup\{x_n, x_{n+1}, \ldots\}\right]$$
$$= \limsup_{n\to\infty} x_n. \tag{11}$$

Note that (10) and (11) imply

$$a_m \leq \liminf_{n\to\infty} x_n \leq \limsup_{n\to\infty} x_n \leq b_m \quad \text{all } m. \tag{12}$$

The following is a useful characterization of the upper and lower limits.

Proposition 3.3. *Let $\{x_n\}$ be a bounded real sequence. The lower limit $\liminf_{n\to\infty} x_n$ is the unique real number a that has the following two properties.*

(a*) *For each $\varepsilon > 0$ there are only finitely many indices n such that $x_n \leq a - \varepsilon$.*
(a**) *For each $\varepsilon > 0$ there are infinitely many indices n such that $x_n < a + \varepsilon$.*

The upper limit $\limsup_{n\to\infty} x_n$ is the unique real number b that has the following two properties.

(b*) *For each $\varepsilon > 0$ there are only finitely many indices n such that $x_n \geq b + \varepsilon$.*
(b**) *For each $\varepsilon > 0$ there are infinitely many indices n such that $x_n > b - \varepsilon$.*

Proof: Suppose first that a has properties (a*) and (a**). Property (a*) implies that no number c *smaller* than a can have property (a**) (choose $\varepsilon = (a-c)/2$, so $a - \varepsilon = c + (a-c)/2$). Property (a**) implies that no number *larger* than a can have property (a*). Therefore there is at most one such number a.

Now take a to be the lower limit of $\{x_n\}$, that is, the least upper bound of the sequence $\{a_n\}$ of (9). Given $\varepsilon > 0$, the smaller number $a - \varepsilon$ is not an upper bound, so there is some N such that

$$a - \varepsilon < a_N = \inf\{x_N, x_{N+1}, \ldots\}.$$

This proves (a*). On the other hand, given an index N, we deduce from the fact that $a_N \leq a < a + \varepsilon$ that there is some index $n \geq N$ such that $x_n < a + \varepsilon$. This proves (a**).

The proof that the upper limit is the unique number that has properties (b*) and (b**) is very similar and is left as an exercise. □

This leads to one criterion for convergence.

Proposition 3.4. *A bounded real sequence $\{x_n\}$ converges if and only if its upper and lower limits are equal.*

Proof: Suppose first that the upper and lower limits are each equal to x. Then, given $\varepsilon > 0$, it follows from Proposition 3.3 that there are only finitely many indices n for which $x_n \leq x - \varepsilon$ or $x_n \geq x + \varepsilon$. Consequently, if N is large enough, $n \geq N$ implies $|x_n - x| < \varepsilon$ and x is the limit.

Conversely, suppose that $\{x_n\}$ converges to x. This means, of course, that, given $\varepsilon > 0$, there is an index N such that $n \geq N$ implies $x - \varepsilon < x_n < x + \varepsilon$. Therefore x has the properties, listed in the statement of Proposition 3.3, that characterize the upper and lower limits. □

Remark. The criterion in Proposition 3.4 is of limited practical interest – one would have to be able to *determine* the upper and lower limits – but it plays a key role in the next section.

Exercises

1. For each of the following real sequences $\{x_n\}_1^\infty$, determine
$$y_n = \inf\{x_n, x_{n+1}, x_{n+2}, \ldots\}, \quad z_n = \sup\{x_n, x_{n+1}, x_{n+2}, \ldots\}.$$
Also determine $\liminf_{n\to\infty} x_n$ and $\limsup_{n\to\infty} x_n$.
   (a) $x_n = (-1)^n + \frac{1}{n}$.
   (b) $x_n = (-1)^n - \frac{1}{n}$.
   (c) $x_n = (-1)^n[1 + \frac{1}{n}]$.
2. Can both sequences $\{a_n\}$ and $\{b_n\}$ of (9) be strictly monotone?
3. Suppose that $\{a_n\}_1^\infty$ and $\{b_n\}_1^\infty$ are bounded real sequences.
   (a) Prove that
$$\limsup_{n\to\infty}(a_n + b_n) \leq \limsup_{n\to\infty} a_n + \limsup_{n\to\infty} b_n.$$
   (b) Show by example that strict inequality can occur.
   (c) Can strict inequality occur if one of the sequences converges?
4. Suppose that $\{a_n\}_1^\infty$ and $\{b_n\}_1^\infty$ are bounded nonnegative sequences. Prove that
$$\limsup_{n\to\infty} a_n b_n \leq \limsup_{n\to\infty} a_n \cdot \limsup_{n\to\infty} b_n.$$
5. Prove the second part of Proposition 3.3.

3C. The Cauchy Criterion

The definition of convergence of a sequence involves its limit. How can one tell whether a series converges if one has not already determined the limit? An answer to this question was given by Cauchy, and the concept he introduced has been named accordingly.

Definition. A complex sequence $\{a_n\}$ is said to be a *Cauchy sequence* if for each $\varepsilon > 0$ there is an index N such that, if the indices n and m are each greater than or equal to N, then $|a_n - a_m| < \varepsilon$.

Remark. It is easy to see that $\{b_n + ic_n\}$ is a Cauchy sequence if and only if the associated real sequences $\{b_n\}$ and $\{c_n\}$ are Cauchy sequences.

Proposition 3.5. *If $\{a_n\}$ converges, then it is a Cauchy sequence.*

Proof: Given $\varepsilon > 0$, choose N so large that $n \geq N$ implies $|a_n - a| < \varepsilon/2$ and use the triangle inequality. □

Proposition 3.6. *If $\{a_n\}$ is a Cauchy sequence, then it is bounded.*

Proof: Choose N so large that $n, m \geq N$ imply $|a_n - a_m| < 1$. In particular, $n \geq N$ implies that $|a_n| = |a_N + (a_n - a_N)| < |a_N| + 1$, and so for every n

$$|a_n| < 1 + \max\{|a_1|, |a_2|, \ldots, |a_N|\}. \quad \square$$

Corollary 3.7. *Every convergent sequence is bounded.*

Now we come to *the* basic result on convergence of complex sequences.

Theorem 3.8: Convergence of Cauchy sequences. *If the complex sequence $\{z_n\}$ is a Cauchy sequence, then it converges.*

Proof: Suppose that $\{z_n\} = \{x_n + iy_n\}$ is a Cauchy sequence. Then the real sequences $\{x_n\}$ and $\{y_n\}$ are Cauchy sequences. If they converge, say to x and to y, then $\{z_n\}$ converges to $x + iy$ (Proposition 3.1). Consequently, it is enough to prove that any *real* Cauchy sequence converges.

Suppose that $\{x_n\}$ is a real Cauchy sequence. Then it is bounded, so it has upper and lower limits a and b. We only need to show that $a = b$. Let $\{a_n\}$ and $\{b_n\}$ be the sequences (9). Given $\varepsilon > 0$, choose N so that $n \geq N$ implies $x_N - \varepsilon < x_n < x_N + \varepsilon$. It follows from this, the definitions of a_n and b_n, and (12) that

$$x_N - \varepsilon \leq a_N \leq a \leq b \leq b_N \leq x_N + \varepsilon$$

and therefore $|b - a| < 2\varepsilon$. This is true for every $\varepsilon > 0$; so $a = b$ and, by Proposition 3.4, the sequence $\{x_n\}$ has limit $x = a$. □

Exercises

1. Suppose that $\{a_n\}$ is a convergent sequence whose terms are positive and nondecreasing. Suppose that $\{b_n\}$ is a complex sequence with the property that, for each n, $|b_{n+1} - b_n| \leq a_{n+1} - a_n$. Prove that $\{b_n\}$ converges.
2. Prove that Corollary 3.13, for real sequences, is equivalent to the Least Upper Bound Property. More precisely, assume all the axioms for the reals from Section 2A except O6, and assume that Corollary 3.13 is valid for real sequences. Show that O6 is a consequence of these assumptions.

3D. Algebraic Properties of Limits

Definition. If $\{a_n\}$ and $\{b_n\}$ are two complex sequences, the *sum, difference, and product* are the sequences with terms $a_n + b_n$, $a_n - b_n$, and $a_n b_n$, respectively. If no term of b_n is zero, then the *quotient* is the sequence with terms a_n/b_n.

Theorem 3.9. *Suppose that $\{a_n\}$ and $\{b_n\}$ are convergent complex sequences with limits a and b, and suppose that c is a complex number. Then*

$$\lim_{n \to \infty} c\, a_n = c a; \tag{13}$$

$$\lim_{n \to \infty} (a_n \pm b_n) = a \pm b; \tag{14}$$

$$\lim_{n \to \infty} (a_n b_n) = ab. \tag{15}$$

Moreover, if $b_n \neq 0$, all n, and $b \neq 0$, then

$$\lim_{n \to \infty} \frac{a_n}{b_n} = \frac{a}{b}. \tag{16}$$

Proof: The assertions (13) and (14) are left as exercises. To prove (15), observe that

$$|a_n b_n - ab| = |(a_n b_n - a b_n) + (a b_n - ab)| \leq |a_n b_n - a b_n| + |a b_n - ab|$$
$$= |a_n - a||b_n| + |a||b_n - b|. \tag{17}$$

We know that if n is large enough, then $|b_n| = |b + (b_n - b)|$ will be less than $|b| + 1$. Given $\varepsilon > 0$ we can choose N so large that $n \geq N$ implies $|b_n| \leq |b| + 1$ and also

$$|a_n - a| < \frac{\varepsilon}{2|b| + 2}, \quad |b_n - b| < \frac{\varepsilon}{2|a| + 2}. \tag{18}$$

It follows that $n \geq N$ implies $|a_n b_n - ab| < \varepsilon$.

The proof of (16) is somewhat similar. Note that if n is large, $|b_n - b| \leq \frac{1}{2}|b|$, which implies that $|b_n| \geq \frac{1}{2}|b|$. For such n,

$$\left|\frac{a_n}{b_n} - \frac{a}{b}\right| = \left|\frac{a_n b - a b_n}{b b_n}\right|$$

$$= \left|\frac{(a_n b - ab) + (ab - a b_n)}{b b_n}\right| \leq \left|\frac{a_n b - ab}{b b_n}\right| + \left|\frac{ab - a b_n}{b b_n}\right|$$

$$= \frac{|b||a_n - a|}{|b_n||b|} + \frac{|a||b_n - b|}{|b||b_n|} \leq |a_n - a|\frac{2}{|b|} + |b_n - b|\frac{|a|}{2|b|^2}.$$

As in the proof of (16), we can control the size of the last expression by choosing n so large that $|a_n - a|$ and $|b_n - b|$ are sufficiently small. $\square$

Remark. Note the trick of breaking the "double" variation $a_n b_n - ab$ into two "single variations" by subtracting and adding ab_n, and the simple modification for dealing with $a_n/b_n - a/b$.

The following result is a partial generalization of (13) that will be used when we come to power series. The proof is left as an exercise.

Proposition 3.10. *If $\{b_n\}$ is a bounded real sequence and $\{a_n\}$ is a positive sequence with limit a, then*

$$\liminf_{n \to \infty} a_n b_n = a \liminf_{n \to \infty} b_n; \quad \limsup_{n \to \infty} a_n b_n = a \limsup_{n \to \infty} b_n. \tag{19}$$

Exercises

1. Prove (13) and (14).
2. Prove Proposition 3.10.
3. Suppose that $\{z_n\}_1^\infty$ is a complex sequence with limit z. Let $\{w_n\}_1^\infty$ be the sequence of arithmetic means $w_n = (z_1 + z_2 + \ldots + z_n)/n$. Prove that $\lim_{n \to \infty} w_n = z$.
4. Suppose that $\lim_{n \to \infty} z_n = z$. Let

$$w_n = \sum_{k=0}^{n} 2^{-n} \binom{n}{k} z_k,$$

   where $\binom{n}{k}$ is the binomial coefficient $n!/k!(n-k)!$. Prove that $\lim_{n \to \infty} w_n = z$.
5. Suppose that $\{z_n\}_1^\infty$ and $\{w_n\}_1^\infty$ are complex sequences with limits z and w, respectively. Show that the following limit exists:

$$\lim_{n \to \infty} \frac{z_1 w_n + z_2 w_{n-1} + z_3 w_{n-2} + \ldots + z_n w_1}{n}.$$

3E. Subsequences

Definition. A *subsequence* of a sequence $\{a_n\}$ is a sequence $\{b_n\}$ whose terms are among the terms of $\{a_n\}$, with increasing indices. In other words, there is a sequence $\{n_k\}_{k=1}^{\infty}$ of positive integers such that

$$n_1 < n_2 < n_3 < \ldots < n_k < \ldots, \qquad b_k = a_{n_k}. \tag{20}$$

(In particular, any sequence is a subsequence of itself, with each $n_k = k$.)

For example, the first sequence (1) has among its subsequences the sequences

$$2, 4, 6, 8, 10, \ldots; \quad 1, 4, 9, 16, 25, \ldots,$$

and the second sequence (1) has among its subsequences the sequences

$$1, 1, 1, 1, 1, 1, \ldots; \quad -1, 0, -1, 0, -1, 0, \ldots.$$

Proposition 3.11. *A complex sequence $\{a_n\}$ converges to a if and only if each of its subsequences converges to a.*

Proof: If $\{a_n\}$ converges to a, it is very easy to prove that each subsequence has limit a. Conversely, $\{a_n\}$ is a subsequence of itself, so the convergence of each subsequence trivially implies the convergence of the sequence $\{a_n\}$ itself. □

For bounded real sequences there is a close relationship between the convergent subsequences and the upper and lower limits of the original sequence, as indicated in the next proposition. The proof is left as an exercise.

Proposition 3.12. *Suppose that $\{x_n\}$ is a bounded real sequence.*

(a) *If $\{y_n\}$ is any convergent subsequence, then*

$$\liminf_{n \to \infty} x_n \leq \lim_{n \to \infty} y_n \leq \limsup_{n \to \infty} x_n.$$

(b) *There is a subsequence that converges to $\liminf_{n \to \infty} x_n$ and also a subsequence that converges to $\limsup_{n \to \infty} x_n$.*

A particular consequence of part (b) is important.

Corollary 3.13: The Bolzano-Weierstrass Theorem. *Each bounded real or complex sequence has a convergent subsequence.*

Proof: The real case follows from Proposition 3.12. For a complex sequence $\{x_n + i y_n\}$, choose a subsequence whose real parts converge and then a subsequence of that one whose imaginary parts converge. □

Exercises

1. Prove Proposition 3.12.
2. Suppose that $\{x_n\}_1^\infty$ is a real sequence, and suppose that for each $k \in \mathbb{N}$ there is a subsequence of $\{x_n\}_1^\infty$ that converges to $1/k$. Prove that there is a subsequence that converges to 0.
3. Suppose that $\{x_n\}_1^\infty$ is a bounded real sequence. For each $m \in \mathbb{N}$, let S_m be the set $\{x_m, x_{m+1}, x_{m+2}, \ldots\}$.
   (a) Suppose that for some m, the set S_m does not contain its least upper bound. Show that $\{x_n\}_1^\infty$ has a strictly increasing subsequence.
   (b) Suppose that for all m, S_m does contain its least upper bound. Show that $\{x_n\}_1^\infty$ has a nonincreasing subsequence. Conclude that every bounded real sequence has a monotone subsequence.
4. Suppose that $\{x_n\}_1^\infty$ is a bounded real sequence and $b = \limsup_{n \to \infty} x_n$. Show that there is a monotone subsequence converging to b.

3F. The Extended Reals and Convergence to $\pm\infty$

It is sometimes convenient to consider the *extended real numbers*; by definition, this is the set that is the union of $\mathbb{R}$ and the two-element set $\{-\infty, +\infty\}$. In this enlarged set one extends the standard algebraic operations and the order relation to include the following:

$$a + (+\infty) = (+\infty) + a = +\infty, \quad \text{all } a \in \mathbb{R}; \tag{21}$$
$$a + (-\infty) = (-\infty) + a = -\infty. \quad \text{all } a \in \mathbb{R}; \tag{22}$$
$$a \cdot (+\infty) = +\infty \quad \text{if } a > 0; \quad = -\infty \quad \text{if } a < 0; \tag{23}$$
$$a \cdot (-\infty) = -\infty \quad \text{if } a > 0; \quad = +\infty \quad \text{if } a < 0; \tag{24}$$
$$a/(\pm\infty) = 0 \quad \text{all } a \in \mathbb{R}; \tag{25}$$
$$(+\infty) + (+\infty) = +\infty; \quad (-\infty) + (-\infty) = -\infty; \tag{26}$$
$$(+\infty) \cdot (+\infty) = +\infty = (-\infty) \cdot (-\infty); \tag{27}$$
$$(+\infty) \cdot (-\infty) = -\infty = (-\infty) \cdot (+\infty). \tag{28}$$

The various field axioms remain valid when each of the expressions that occur is defined, but $0 \cdot (\pm\infty)$, $(\pm\infty) \cdot 0$, $(+\infty) + (-\infty)$, and $(-\infty) + (+\infty)$ are *not defined*. [There is a reason why they are not defined; we return to this point shortly. In particular, do not yield, yet, to the temptation to set $0 \cdot (\pm\infty)$ equal to 0.]

Definition. A real sequence $\{x_n\}$ is said to have limit $+\infty$ if for each real M there is a positive integer N such that $n \geq N$ implies $x_n \geq M$. Similarly, $\{x_n\}$ is said to have limit $-\infty$ if for each real L there is $N \in \mathbb{N}$ such that $n \geq N$

implies $x_n \leq L$. In these cases we write $\lim_{n \to \infty} x_n = +\infty$ or $\lim_{n \to \infty} x_n = -\infty$, respectively.

For example, the first sequence in (1) has limit $+\infty$. The various parts of Theorem 3.9 remain valid for real sequences when some or all of the limits are $\pm\infty$, Some of these are included among the exercises. To see what might go wrong, consider the sequences with terms

$$a_n = 0, \quad b_n = 1/n, \quad c_n = n, \quad d_n = n^2, \quad n \in \mathbb{N}.$$

Then $\{a_n\}$ and $\{b_n\}$ both have limit 0 and $\{c_n\}$ and $\{d_n\}$ both have limit $+\infty$, while $\{a_n c_n\}$ and $\{a_n d_n\}$ have limit 0, $\{b_n c_n\}$ has limit 1, and $\{b_n d_n\}$ has limit $+\infty$. Thus there is no way we can define $0 \cdot (+\infty)$ to make the limit of the product equal to the product of the limits in all these examples.

Note that *a nondecreasing real sequence either has a finite limit or has limit $+\infty$*. The proof is left as an exercise.

Remark. The plus sign is often omitted when writing $+\infty$.

Exercises

1. Suppose that all x_n's are positive. Prove that $\{x_n\}_1^\infty$ has limit ∞ if and only if $\{1/x_n\}_1^\infty$ has limit 0.
2. Prove that any nondecreasing real sequence has a limit, allowing $+\infty$ as a possible limit.
3. Prove (14) and (15) when $\{a_n\}$ and $\{b_n\}$ are real sequences and $\{a_n\}$ has limit $a \in \mathbb{R}$, $a > 0$, while $\{b_n\}$ has limit $+\infty$.
4. Suppose that $\{x_n\}_1^\infty$ is a sequence of positive reals with limit $x \geq 0$ (allowing $x = +\infty$). Prove that $\lim_{n \to \infty} (x_1 x_2 \cdots x_n)^{1/n} = x$.
5. Suppose that $\{z_n\}_1^\infty$ is a complex sequence with limit z and suppose that $\{a_n\}_1^\infty$ is a positive sequence such that $\lim_{n \to \infty} (a_1 + a_2 + \ldots + a_n) = +\infty$. Prove that

$$\lim_{n \to \infty} \frac{a_1 z_1 + a_2 z_2 + \ldots a_n z_n}{a_1 + a_2 + \ldots + a_n} = z.$$

6. Prove
   (a) $\lim_{n \to \infty} x^n / n^k = +\infty$ if $x > 1$ and $k \in \mathbb{N}$;
   (b) $\lim_{n \to \infty} n^k x^n = 0$ if $|x| < 1$ and $k \in \mathbb{N}$;
   (c) $\lim_{n \to \infty} x^{1/n} = 1$ if $x > 0$.
7. Use the preceding and/or succeeding exercise to rank the following in order of size for very large positive integers n; you may use the fact from the next section that $\lim_{n \to \infty} (1 + \frac{1}{n})^n = e = 2.718\ldots$:

$$n! \quad n^{100} \quad n^{n/2} \quad 2^n \quad \left(\frac{n}{2}\right)^n \quad n^n.$$

8. Suppose that a_n, and b_n are positive, all $n \in \mathbb{N}$. Suppose that there are positive constants r and N with $r < 1$ such that

$$\frac{a_{n+1}}{a_n} \leq r \frac{b_{n+1}}{b_n} \quad \text{if } n \geq N.$$

Show that $\lim_{n\to\infty} a_n/b_n = 0$.

9. Suppose that $\{x_n\}_1^\infty$ is a real sequence such that $x_{n+m} \leq x_n + x_m$ for each pair of indices n and m. Prove that either the sequence $\{x/n\}_{n=1}^\infty$ converges or else $\lim_{n\to\infty} \frac{x_n}{n} = -\infty$.

3G. Sizes of Things: The Logarithm

The derivative, integration, the Fundamental Theorem of Calculus, the exponential, and the (natural) logarithm are all treated rigorously in Chapter 8, but they are not likely to be new to any reader of this book. To add to our repertoire in treating limits and in comparing the sizes of things, it is very useful to jump ahead and to use various facts about the logarithm.

We write $\log x$ for the natural logarithm (base e), which is often written $\ln x$. It is defined for all positive x and has the properties

$$\log 1 = 0; \quad \frac{d}{dx}\log x = \frac{1}{x}; \quad e^{\log x} = x. \tag{29}$$

It follows from the first two of these properties that

$$\log x = \int_1^x \frac{dt}{t}. \tag{30}$$

It will be useful to have a good estimate of $\log x$ when x is close to 1. We take $x = 1 + h$ and assume that h is positive. Then (30) becomes

$$\log(1 + h) = \int_1^{1+h} \frac{dt}{t}. \tag{31}$$

The interval of integration has length h, and on this interval the integrand is between $1/(1+h)$ and 1 (see Figure 2). Therefore,

$$\frac{h}{1+h} < \log(1+h) < h, \quad \text{if } h > 0. \tag{32}$$

It follows from (32) that

$$\frac{a}{n+a} < \log\left(1 + \frac{a}{n}\right) < \frac{a}{n}, \quad \text{if } a \geq 0;$$

$$\lim_{n\to\infty} n \log\left(1 + \frac{a}{n}\right) = a. \tag{33}$$

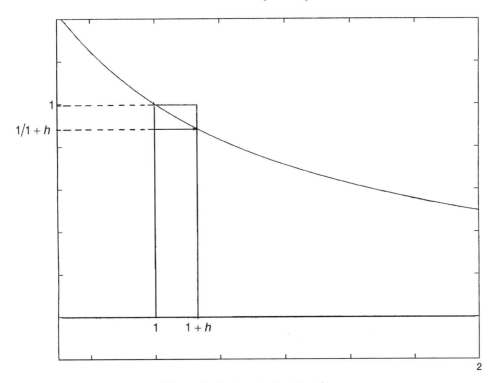

Figure 2. Estimating $\log(1 + h)$.

Exponentiating, we get

$$e^{na/(n+a)} < \left(1 + \frac{a}{n}\right)^n < e^a;$$
$$\lim_{n \to \infty} \left(1 + \frac{a}{n}\right)^n = e^a. \tag{34}$$

Exercise

1. Prove that $\lim_{n \to \infty} n \left[n^{1/n} - 1\right] = +\infty$.

Additional Exercises for Chapter 3

1. Suppose that $F(x) = ax(1 - x)$, where $0 \leq a \leq 4$.
   (a) Show that if $x \in I$, where I is the unit interval $[0, 1]$, then $F(x) \in I$. This means that we can define sequences in I by choosing any $x_1 \in I$ and letting $x_{n+1} = F(x_n)$, $n \geq 1$.
   (b) Prove that if $0 \leq a \leq 1$, then, for any choice of $x_1 \in I$, $\lim_{n \to \infty} x_n = 0$.
   (c) Prove that if $1 < a \leq 3$, then there are exactly two points y in I such that $F(y) = y$, and for any $0 < x_1 \leq 1$ the sequence converges to the larger of these two points.

(d) Prove that for $3 < a \leq 4$ there are two points as in (c) and also two distinct points p and q in I such that $F(p) = q$ and $F(q) = p$. Show that for $3 < a < 1 + \sqrt{6}$ and any x_1 in I, either the sequence stays at one of the two fixed points of (c), or else its even terms converge to one of the points p or q and its odd terms converge to the other.

Exercises 2–6 deal with the *Fibonacci sequence*. This is the sequence $\{F_n\}_1^\infty$ that is characterized by the equation (*) $F_{n+2} = F_n + F_{n+1}$ together with the starting conditions $F_1 = F_2 = 1$.

2. For what values of r does the sequence $\{r^n\}_{n=1}^\infty$ satisfy the equation (*)?
3. Show that if sequences $\{x_n\}_1^\infty$ and $\{y_n\}_1^\infty$ both satisfy the equation (*) and a and b are constants, then the sequence $\{c_n\}_1^\infty$ with terms $c_n = ax_n + by_n$ also satisfies (*).
4. Show that there are constants $a, b, r,$ and s such that $F_n = ar^n + bs^n$ for every $n \in \mathbb{N}$, while $r > 1$ and $|s| < 1$.
5. Compare F_n in size to $(8/5)^n$ and $(13/8)^n$. Compute $\lim_{n \to \infty} F_{n+1}/F_n$.
6. Are $17/12$ and $41/29$ good approximations to $\sqrt{2}$? Discuss the reason for this in connection with the sequence $\{G_n\}_1^\infty$ defined by $G_1 = G_2 = 1$, $G_{n+2} = G_{n+1} + \frac{1}{4}G_n$.

Exercises 7–13 deal with the *Mandelbrot set*. This is the set M of all complex numbers c with the property that the sequence of complex numbers $\{z_n\}_0^\infty$ defined as follows is *bounded*: $z_0 = 0$, $z_{n+1} = z_n^2 + c$. (See almost any book or article on "fractals.")

7. Compute the sequence for each of the choices: $c = 0, c = -1, c = i, c = -i, c = -2$.
8. Which of the following is in M: $\frac{1}{4}, \frac{1}{3}, -\frac{1}{3}, 1 + i$?
9. Prove that if c is positive, then the sequence is strictly increasing. Assuming that the limit is finite, compute it as a function of c.
10. Prove that $M \cap [0, \infty) = [0, \frac{1}{4}]$.
11. Prove that $M \cap (-\infty, 0] = [-2, 0]$.
12. Prove that if $c \in \mathbb{C}$ and $|c| > 2$, then $c \notin M$.
13. Prove that if for some $m \in \mathbb{N}$, $|z_m| > 2$, then $c \notin M$.

4
Series

The principal concepts and techniques of infinite series are introduced here, including the standard convergence tests and the difference between absolute and conditional convergence. Using Euler's constant, we evaluate some of the series introduced in the first chapter.

4A. Convergence and Absolute Convergence

If $\{a_n\}$ is a complex sequence, we associate to it the *series* denoted by $\sum_1^\infty a_n$. For the moment this is just a formal expression. To give it meaning, we look at the sequence $\{s_n\}$ of *partial sums*

$$s_1 = a_1, \quad s_2 = a_1 + a_2, \quad \ldots, \quad s_n = \sum_{k=1}^n a_k. \tag{1}$$

The series $\sum_1^\infty a_n$ is said to be *convergent*, or to *converge*, if the sequence of partial sums converges. If $\{s_n\}$ has limit s, then s is said to be the *sum* of the series $\sum_1^\infty a_n$ and one writes

$$\sum_{n=1}^\infty a_n = \lim_{n \to \infty} \sum_{k=1}^n a_k = s. \tag{2}$$

Otherwise the series is said to *diverge* or to be *divergent*.

Remarks

1. For a convergent series the expression $\sum_1^\infty a_n$ does double duty. It stands for the series itself (with *terms* a_n and partial sums (1)) and also for the complex number that is its sum, depending on the context.

2. Consider $a_n = b_n + ic_n$. The real parts of the partial sums are the partial sums of the real series $\sum_1^\infty b_n$, and so on, so $\sum_1^\infty a_n$ converges if and only if both $\sum_1^\infty b_n$ and $\sum_1^\infty c_n$ converge (Proposition 3.1).
3. The terms of a series can be recovered from the partial sums:

$$a_1 = s_1, \quad a_n = s_n - s_{n-1} \quad \text{if } n > 1. \tag{3}$$

Therefore, for a series $\sum_1^\infty a_n$ to be convergent, it is necessary that $\lim_{n\to\infty} a_n = 0$. We shall see later that this condition is not sufficient.

4. We also deal frequently with series whose terms are indexed by the nonnegative integers $0, 1, 2, \ldots$. This requires minor modifications in the notation and in (1)–(3).

The simplest examples of convergent series are the *geometric series*.

Theorem 4.1: Geometric series. *Let z be a complex number. The series $\sum_0^\infty z^n$ converges if and only if $|z| < 1$. If so, the sum is $1/(1-z)$.*

Proof: If $|z| \geq 1$, then the terms of the series do not have limit 0 and so the series diverges.

Suppose that $|z| < 1$. The partial sums are

$$s_n = \sum_{k=0}^n z^n = \frac{1 - z^{n+1}}{1 - z}$$

(equation (5) of Chapter 1) and $\lim_{n\to\infty} z^{n+1} = 0$, so the limit is $1/(1-z)$. □

In principle, the theory of series is contained in the theory of sequences: One looks at the sequence of partial sums. We have not yet developed the theory of sequences far enough to include much useful information about series, but we can obtain some algebraic properties from Theorem 3.9.

Theorem 4.2. *Suppose that $\sum_1^\infty a_n$ and $\sum_1^\infty b_n$ are convergent and c is a complex number. Then the series $\sum_1^\infty ca_n$, $\sum_1^\infty (a_n + b_n)$, and $\sum_1^\infty (a_n - b_n)$ are convergent and*

$$\sum_{n=1}^\infty ca_n = c \sum_{n=1}^\infty a_n; \quad \sum_{n=1}^\infty (a_n \pm b_n) = \sum_{n=1}^\infty a_n \pm \sum_{n=1}^\infty b_n. \tag{4}$$

Definition. The series $\sum_1^\infty a_n$ is said to be *absolutely convergent*, or to *converge absolutely*, if the real series $\sum_1^\infty |a_n|$ converges.

4A. Convergence and Absolute Convergence

This definition would be unfortunate if there were a series that was absolutely convergent but not convergent, since the former sounds like a stronger condition. In fact it is a stronger condition.

Theorem 4.3: Absolute convergence. *If the series $\sum_1^\infty a_n$ is absolutely convergent, then it is convergent and the sum satisfies*

$$\left| \sum_{n=1}^\infty a_n \right| \leq \sum_{n=1}^\infty |a_n|. \tag{5}$$

Proof: Let $\{s_n\}$ be the sequence of partial sums of $\sum_1^\infty a_n$ and let $\{t_n\}$ be the sequence of partial sums of $\sum_1^\infty |a_n|$. To show convergence, we show that $\{s_n\}$ is a Cauchy sequence. In fact, if $m < n$, then

$$|s_n - s_m| = |a_{m+1} + a_{m+2} + \ldots + a_n|$$
$$\leq |a_{m+1}| + |a_{m+2}| + \ldots + |a_n| = t_n - t_m = |t_n - t_m|. \tag{6}$$

But $\{t_n\}$ is a Cauchy sequence, so (6) is small if n and m are large.

The inequality (5) follows from the inequalities $|s_n| \leq t_n$ for the partial sums. □

Notice that the terms of $\sum_1^\infty |a_n|$ are nonnegative. Thus the following theorem specializes to give a necessary and sufficient condition for absolute convergence.

Theorem 4.4: Series with nonnegative terms. *If $b_n \geq 0$ for all n, then $\sum_1^\infty b_n$ converges if and only if the sequence of partial sums is a bounded sequence.*

Proof: The sequence of partial sums in this case is nondecreasing, so it has a (finite) limit if and only if it is bounded (Theorem 3.2). □

If the terms b_n are nonnegative, then divergence of $\sum_1^\infty b_n$ means that the partial sums have limit $+\infty$. In this case one writes

$$\sum_{n=1}^\infty b_n = \infty.$$

It is important to remember that this notation presupposes that all b_n's are nonnegative.

Exercises

1. Suppose that $\{s_n\}_1^\infty$ is the sequence of partial sums of the series $\sum a_n$. Suppose that $\lim_{n\to\infty} a_n = 0$ and that $\lim_{n\to\infty} s_{2n}$ exists. Prove that the series converges.
2. Find the sum of the series $\sum_1^\infty 1/n(n+1)$.
3. Suppose that $a_1 \geq a_2 \geq a_3 \geq \ldots$, and suppose that $\lim_{n\to\infty} a_n = 0$. Show that if $\sum_1^\infty a_n$ converges, then $\lim_{n\to\infty} na_n = 0$. (This is easy if you know that $\lim_{n\to\infty} na_n$ exists, but why should it exist?)

4B. Tests for (Absolute) Convergence

Theorem 4.4 is the basis for the first general test for convergence.

Theorem 4.5: Comparison Test. *Suppose that $b_n \geq 0$ for all n, and suppose that there are constants M and N such that*

$$|a_n| \leq M b_n, \quad \text{all } n \geq N \tag{7}$$

and $\sum_1^\infty b_n$ converges. Then $\sum_1^\infty a_n$ converges.

Proof: It follows from the assumptions that the partial sums of $\sum_1^\infty b_n$ are bounded, and (7) can be used to show easily that the partial sums of $\sum_1^\infty |a_n|$ are bounded. Therefore $\sum_1^\infty a_n$ is absolutely convergent, and hence convergent. □

The Comparison Test is not of much use without a stock of convergent series available for comparison. The only nontrivial positive series that we have, so far, are the geometric series $\sum r^n$, with $0 < r < 1$. The next two tests take advantage of these series.

Theorem 4.6: Ratio Test. *Suppose that $a_n \neq 0$ for all n. Then*

(a) $\quad \limsup_{n\to\infty} \dfrac{|a_{n+1}|}{|a_n|} < 1 \quad \text{implies that} \quad \sum_{n=1}^\infty a_n \text{ converges;}$ absolutely

(b) $\quad \dfrac{|a_{n+1}|}{|a_n|} \geq 1 \quad \text{for } n \geq N \quad \text{implies that} \quad \sum_{n=1}^\infty a_n \text{ diverges.}$

In particular, if the limit $L = \lim_{n\to\infty} |a_{n+1}|/|a_n|$ exists, then the series converges if $L < 1$ and diverges if $L > 1$.

Proof: Suppose that the condition in (a) is satisfied and choose r so that

$$\limsup_{n\to\infty} \frac{|a_{n+1}|}{|a_n|} < r < 1.$$

Only finitely many of the quotients can exceed r (Proposition 3.3), so there is $N \in \mathbb{N}$ such that the quotient is $\leq r$ for $n \geq N$. This means that, for each $n > N$,

$$|a_n| \leq r|a_{n-1}| \leq r(r|a_{n-2}|) = r^2|a_{n-2}| \leq \ldots \leq r^{n-N}|a_N| = M r^n,$$
$$M = r^{-N}|a_N|.$$

The Comparison Test and the convergence of $\sum_0^\infty r^n$ imply convergence.

Now suppose that the condition in (b) is satisfied. Then $|a_n| \geq |a_N|$ for $n \geq N$, so the terms of the series do not have limit 0. □

Examples

1. $\sum_0^\infty 1/n!$. The ratio is $1/(n+1)$, so the series converges.
2. $\sum_1^\infty 1/n$. The ratio is $n/(n+1) = 1/(1+1/n)$, which is < 1 but has limit 1, so the ratio test is inconclusive. The same is true for $\sum_1^\infty 1/n^2$.

The next test is usually harder to apply than the Ratio Test, but it is important, in principle, for power series.

Theorem 4.7: Root Test

(a) $$\limsup_{n \to \infty} |a_n|^{1/n} < 1 \quad \text{implies that} \quad \sum_{n=1}^\infty a_n \text{ converges;}$$

(b) $$\limsup_{n \to \infty} |a_n|^{1/n} > 1 \quad \text{implies that} \quad \sum_{n=1}^\infty a_n \text{ diverges.}$$

In particular, if the limit $L = \lim_{n \to \infty} |a|^{1/n}$ exists, then $\sum_1^\infty a_n$ converges if $L < 1$ and diverges if $L > 1$.

Proof: If the condition in (a) is satisfied, choose r such that

$$\limsup_{n \to \infty} |a_n|^{1/n} < r < 1.$$

By Proposition 3.3 once more, this implies that $|a_n|^{1/n} \leq r$ for $n \geq N$. Thus $|a_n| \leq r^n$ for $n \geq N$, and the Comparison Test implies convergence.

If the condition in (b) is satisfied, then $|a_n| \geq 1$ for infinitely many values of n, by Proposition 3.3, so the terms do not have limit 0. □

Example. $\sum_1^\infty 1/n$. By Exercise 5 of Section 3F, $\lim_{n \to \infty} n^{1/n} = 1$, so the Root Test is inconclusive. The same is true for the series $\sum_1^\infty 1/n^2$.

Neither the Ratio Test nor the Root Test tells us whether $\sum_1^\infty 1/n$ or $\sum_1^\infty 1/n^2$ converges or diverges. The terms of these series are positive, so all we need determine is whether or not the partial sums are bounded. For large m,

$$1 + \frac{1}{2} + \frac{1}{3} + \frac{1}{4} + \ldots + \frac{1}{2^m}$$
$$= 1 + \frac{1}{2} + \left(\frac{1}{3} + \frac{1}{4}\right) + \left(\frac{1}{5} + \frac{1}{6} + \frac{1}{7} + \frac{1}{8}\right) + \ldots + \left(\frac{1}{2^{m-1}+1} + \ldots + \frac{1}{2^m}\right)$$
$$> 1 + \frac{1}{2} + \frac{1}{2} + \frac{1}{2} + \ldots + \frac{1}{2} = 1 + \frac{m}{2}.$$

Therefore, the partial sums are unbounded and the *harmonic series* diverges:

$$\sum_1^\infty \frac{1}{n} = \infty, \tag{8}$$

On the other hand,

$$1 + \frac{1}{4} + \frac{1}{9} + \frac{1}{16} + \ldots + \frac{1}{(2^{m+1}-1)^2}$$
$$= 1 + \left(\frac{1}{4} + \frac{1}{9}\right) + \left(\frac{1}{16} + \frac{1}{25} + \frac{1}{36} + \frac{1}{49}\right)$$
$$+ \ldots + \left(\frac{1}{2^{2m}} + \ldots + \frac{1}{(2^{m+1}-1)^2}\right)$$
$$< 1 + \frac{2}{2^2} + \frac{4}{4^2} + \frac{8}{8^2} + \ldots + \frac{2^m}{(2^m)^2} < \sum_{m=0}^\infty \frac{1}{2^m} = 2,$$

so the partial sums are bounded and the series $\sum_1^\infty 1/n^2$ converges.

The two procedures just outlined can be used to prove the next theorem.

Theorem 4.8: 2^m test. *Suppose that the real sequence $\{a_m\}$ is nonincreasing and has limit 0:*

$$a_1 \geq a_2 \geq a_3 \geq \ldots \geq a_n \geq 0; \qquad \lim_{n \to \infty} a_n = 0.$$

Then $\sum_1^\infty a_n$ converges if and only if the series $\sum_{m=0}^\infty 2^m a_{2^m}$ converges.

Example. Consider $\sum_1^\infty 1/n^b$, where $b > 0$. Then $a_n = n^{-b}$, so

$$\sum_{m=0}^\infty 2^m a_{2^m} = \sum_{m=0}^\infty \frac{2^m}{(2^m)^b} = \sum_{m=0}^\infty \left(\frac{1}{2^{b-1}}\right)^m.$$

4B. Tests for (Absolute) Convergence

The last series is a geometric series and it converges if and only if $2^{b-1} > 1$, that is, $b > 1$. In particular, this shows again that $\sum_{1}^{\infty} 1/n$ diverges while $\sum_{1}^{\infty} 1/n^2$ converges.

One good proof deserves another. Let us take a second look at the series $\sum_{1}^{\infty} 1/n^b$ by picturing a sequence of rectangles in the plane. Each rectangle has its base on the horizontal axis, identified with the real line. The n-th rectangle has as its base either the interval $[n-1, n]$ or the interval $[n, n+1]$ and has height $1/n^b$. This picture demonstrates the inequalities

$$\int_{1}^{n+1} \frac{dx}{x^b} \leq \sum_{k=1}^{n} \frac{1}{k^b} \leq 1 + \int_{1}^{n} \frac{dx}{x^b}. \tag{9}$$

Performing the integration, one sees once again that the partial sums of $\sum_{1}^{\infty} 1/n^b$ are bounded if and only if $b > 1$.

The preceding argument can also be put into the form of a general convergence test. (See Figure 3.)

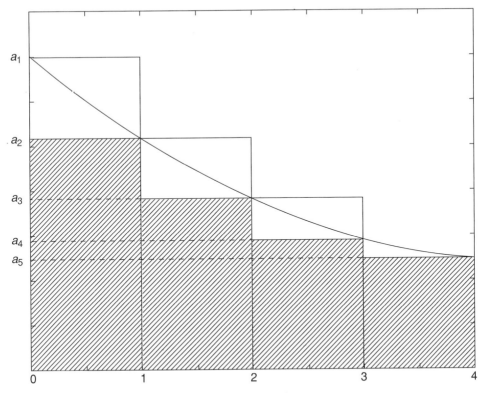

Figure 3. The integral test.

Theorem 4.9: Integral test. *Suppose that f is a nonnegative nonincreasing continuous function defined for $x \geq 1$ and suppose that $a_n = f(n)$ for all $n \in \mathbb{N}$. Then $\sum_1^\infty a_n$ converges if and only if the improper integral $\int_1^\infty f(x)\,dx$ is finite.*

[The cautious and logically minded reader will note that we have not defined several of the terms used here and do not do so until later sections. The good news is that no use is made of Theorem 4.9 in any of the developments leading up to those sections, so the reasoning is not circular. On the other hand, Theorem 4.9 is a beautiful reminder of the analogy between summation and integration – and it may be helpful in some of the exercises.]

Some General Remarks About Deciding Convergence or Divergence

No convergence test has all the answers. For series with positive terms, it is vital to know how fast the terms go to zero. Remember that, for comparison, an inequality

$$|a_n| \leq \frac{1000000}{n^2}$$

is good enough for convergence, and an inequality

$$a_n \geq \frac{1}{1000000\,n}$$

is good enough for divergence. In examining the terms of a series, note what is important and what is not for large n: If an expression like $2^n - n^2$ appears in a numerator or denominator, 2^n is important and n^2 is not (why?). Sometimes simply clearing out the unimportant terms will give the answer. For example, in the case of $n^2 - n$, one can use the inequalities

$$\frac{n^2}{2} \leq n^2 - n \leq n^2 \qquad \text{for } n \geq 2$$

to replace $n^2 - n$ in a denominator by the simpler term n^2.

Exercises

In Exercises 1–18, determine whether the series converges or diverges.

1. $\displaystyle\sum_{n=1}^{\infty} \frac{n^3}{3^n}.$

2. $\displaystyle\sum_{n=1}^{\infty} \frac{(n!)^2}{(2n)!}.$

3. $\sum_{n=1}^{\infty} \frac{(n!)^2 (15)^{n/2}}{(2n)!}.$

4. $\sum_{n=1}^{\infty} \frac{(n-\sqrt{n})}{(n^2+5n)}.$

5. $\sum_{n=1}^{\infty} \frac{\sqrt{n+1}-\sqrt{n}}{n}.$

6. $\sum_{n=1}^{\infty} \frac{2^n n!}{n^n}.$

7. $\sum_{n=1}^{\infty} \frac{3^n n!}{n^n}.$

8. $\sum_{n=1}^{\infty} \frac{1 \cdot 3 \cdot 5 \cdots (2n-1)}{2 \cdot 4 \cdot 6 \cdots 2n}.$

9. $\sum_{n=1}^{\infty} \frac{2 \cdot 4 \cdot 6 \cdots 2n}{1 \cdot 3 \cdot 5 \cdots (2n+1)}.$

10. $\sum_{n=1}^{\infty} (\log n)^{-\log n}.$

11. $\sum_{n=1}^{\infty} (\log n)^{-\log \log n}.$

12. $\sum_{n=1}^{\infty} \left(1 - \frac{\log n}{\log(n+1)}\right).$

13. $\sum_{n=1}^{\infty} \frac{1}{n^{1+1/n}}.$

14. $\sum_{n=1}^{\infty} \left(n^{1/n} - 1\right).$

15. $\sum_{n=1}^{\infty} \sqrt{n}\left(n^{1/n^2} - 1\right).$

16. $\sum_{n=1}^{\infty} (1-a)\left(1-\frac{a}{2}\right)\left(1-\frac{a}{3}\right) \cdots \left(1-\frac{a}{n}\right), \quad a > 0.$

17. $\sum_{n=1}^{\infty} \frac{a(a+1)(a+2) \cdots (a+n-1)}{b(b+1)(b+2) \cdots (b+n-1)}, \quad a, b > 0.$

18. $\sum_{n=1}^{\infty} \frac{a(a+1) \cdots (a+n-1) \cdot c(c+1) \cdots (c+n-1)}{b(b+1) \cdots (b+n-1) \cdot d(d+1) \cdots (d+n-1)}, \quad a, b, c, d > 0.$

19. For what, if any, values of $a > 0$ do the following series converge?

(a) $\sum_{n=2}^{\infty} 1/n(\log n)^a.$ (b) $\sum_{n=3}^{\infty} 1/n \log n (\log \log n)^a.$

20. For what, if any, values of the real numbers a and b do the following series converge?

(a) $\sum_{n=1}^{\infty} \left(\dfrac{1}{n} + \dfrac{a}{n+1} \right)$. (b) $\sum_{n=1}^{\infty} \left(\dfrac{1}{n} + \dfrac{a}{n+1} + \dfrac{b}{n+2} \right)$.

21. Do the following two series converge or diverge?

(a) $1 - \dfrac{1}{\sqrt{2}} + \dfrac{1}{\sqrt{3}} - \dfrac{1}{\sqrt{4}} + \dfrac{1}{\sqrt{5}} - \dfrac{1}{\sqrt{6}} + \cdots$.

(b) $1 + \dfrac{1}{\sqrt{3}} - \dfrac{1}{\sqrt{2}} + \dfrac{1}{\sqrt{5}} + \dfrac{1}{\sqrt{7}} - \dfrac{1}{\sqrt{4}} + \cdots$.

22. Does the following series converge or diverge?

$$1 + \dfrac{1}{3^2} - \dfrac{1}{2} + \dfrac{1}{5^2} + \dfrac{1}{7^2} - \dfrac{1}{4} + \cdots + \dfrac{1}{(4n+1)^2} + \dfrac{1}{(4n+3)^2} - \dfrac{1}{2n+2} + \cdots.$$

23. Is there any real constant a such that the following series converges?

$$1 + \dfrac{1}{\sqrt{3}} - \dfrac{a}{\sqrt{2}} + \dfrac{1}{\sqrt{5}} + \dfrac{1}{\sqrt{7}} - \dfrac{a}{\sqrt{4}} + \cdots.$$

24. Prove Theorem 4.8, the 2^m test.
25. Prove Theorem 4.9, the integral test. (Assume the standard properties of the definite integral.)
26. Let M be the subset of $\mathbb{N}$ consisting of those integers n such that no digit of n in its decimal expression is 5. Prove that $\sum_{n \in M} 1/n$ converges.

4C*. Conditional Convergence

A series that converges but that is not absolutely convergent is said to be *conditionally convergent*, or to *converge conditionally*.

The alternating harmonic series $\sum_{1}^{\infty} (-1)^{n-1}/n$ that we saw in Section 1A is not absolutely convergent. On the other hand, look at the grouping of terms there:

$$1 - \left(\dfrac{1}{2} - \dfrac{1}{3}\right) - \left(\dfrac{1}{4} - \dfrac{1}{5}\right) - \left(\dfrac{1}{6} - \dfrac{1}{7}\right) - \left(\dfrac{1}{8} - \dfrac{1}{9}\right) - \cdots \quad (10)$$

$$= 1 - \dfrac{1}{2 \cdot 3} - \dfrac{1}{4 \cdot 5} - \dfrac{1}{6 \cdot 7} - \dfrac{1}{8 \cdot 9} - \cdots.$$

The regrouped series converges by comparison to $\sum_{1}^{\infty} 1/n^2$, and this suggests that the original series (10) converges. In fact, it shows that the odd partial sums $\{s_{2n+1}\}$ converge; and since the terms have limit 0, it is not difficult to show that the full sequence of partial sums converges and has sum < 1. Thus the series (10) is conditionally convergent.

Similarly, the regrouping

$$1 + \frac{1}{3} - \frac{1}{2} + \frac{1}{5} + \frac{1}{7} - \frac{1}{4} + \frac{1}{9} + \frac{1}{11} - \frac{1}{6} + \cdots$$

$$= 1 + \left(\frac{1}{3} - \frac{1}{2} + \frac{1}{5}\right) + \left(\frac{1}{7} - \frac{1}{4} + \frac{1}{9}\right) + \left(\frac{1}{11} - \frac{1}{6} + \frac{1}{13}\right) + \cdots$$

$$= 1 + \frac{1}{2 \cdot 3 \cdot 5} + \frac{1}{4 \cdot 7 \cdot 9} + \frac{1}{6 \cdot 11 \cdot 13} + \cdots \qquad (11)$$

shows that the last series converges (by comparison with $\sum_1^\infty 1/n^3$); one can conclude first that the subsequence $\{t_{3n+1}\}$ of the sequence of partial sums of (11) converges and then that the series (11) itself converges. Moreover, its sum is > 1. In other words, we have recovered the paradox of Chapter 1 and have shown that indeed these two (infinite) sums are different even though the summands are the same. We return to this question in a moment, after turning to the single most useful criterion for conditional convergence.

Theorem 4.10: Alternating series. *Suppose that the real sequence $\{a_n\}$ is nonincreasing and has limit 0:*

$$a_1 \geq a_2 \geq a_3 \geq \ldots \geq a_n \geq 0; \qquad \lim_{n \to \infty} a_n = 0. \qquad (12)$$

Then the alternating series $\sum_1^\infty (-1)^{n-1} a_n$ converges. Moreover, if $\{s_n\}$ is the sequence of partial sums, then

$$s_{2n} \leq \sum_{n=1}^\infty (-1)^{n-1} a_n \leq s_{2n-1} \qquad \text{all } n. \qquad (13)$$

Proof: It is easy to see by induction that

$$s_2 \leq s_4 \leq \cdots \leq s_{2n} \leq s_{2n-1} \leq \cdots \leq s_3 \leq s_1 \qquad (14)$$

for all n. It follows that the upper and lower limits of $\{s_n\}$ lie between s_{2n} and s_{2n-1} for every n. But $s_{2n-1} - s_{2n} = a_{2n}$ has limit 0. Therefore, the upper and lower limits are the same, and the sequence converges. The inequality (13) follows from (14). □

Let us return to the paradox concerning the series (10) and (11). Note that (11) is a rearrangement of (10), in the following sense.

Definition. A series $\sum_1^\infty b_n$ is said to be a *rearrangement* of the series $\sum_1^\infty a_n$ if there is a 1–1 mapping of $\mathbb{N}$ onto itself, $n \to \sigma(n)$, such that $b_n = a_{\sigma(n)}$, all n. In other words, each index m occurs exactly once among the integers $\sigma(n)$, so each indexed term a_m occurs exactly once as an indexed term b_n.

Absolutely and conditionally convergent series have completely different behaviors under rearrangement.

Theorem 4.11. *Suppose that $\sum_1^\infty a_n$ is absolutely convergent and that $\sum_1^\infty b_n$ is a rearrangement. Then $\sum_1^\infty b_n$ converges and has the same sum as $\sum_1^\infty a_n$.*

Proof: Let $\{s_n\}$ be the sequence of partial sums of $\sum_1^\infty a_n$ and let s be the limit. Let $\{t_n\}$ be the sequence of partial sums of $\sum_1^\infty b_n$. Given $\varepsilon > 0$, choose M so large that

$$\sum_{n=M+1}^\infty |a_n| < \varepsilon/2. \tag{15}$$

It follows from this that $|s_M - s| < \varepsilon/2$. Choose N so large that every one of the first M terms of $\{a_n\}$ occurs among the first N terms of $\{b_n\}$. It follows that for $n \geq N$ the difference $|t_n - s_M|$ is no larger than the left-hand side of (15), and therefore

$$|t_n - s| \leq |t_n - s_M| + |s_M - s| < \varepsilon/2 + \varepsilon/2 = \varepsilon. \quad \square$$

The next theorem shows that it is not at all remarkable that the series (10) and (11) have different limits.

Theorem 4.12: Riemann's Rearrangement Theorem. *Suppose that $\sum_1^\infty a_n$ is a conditionally convergent real series. For each real number s, there is a rearrangement of $\sum_1^\infty a_n$ that converges and has sum s.*

Proof: Let $p_1, p_2, \ldots$ and $q_1, q_2, \ldots$ enumerate, in order, the positive and negative terms, respectively, of $\sum_1^\infty a_n$. The nonnegative series $\sum_1^\infty p_n$ and $\sum_1^\infty (-q_n)$ both diverge. In fact, if both were convergent, it would follow that $\sum_1^\infty |a_n|$ converges, that is, $\sum_1^\infty a_n$ would be absolutely convergent. On the other hand, if one of these series converged and the other diverged, it would follow that the partial sums of $\sum_1^\infty a_n$ diverge either to $+\infty$ or to $-\infty$. The convergence of $\sum_1^\infty a_n$ itself implies that $\{p_n\}$ and $\{q_n\}$ have limit 0. With these facts in mind we can rearrange $\sum_1^\infty a_n$ to sum to s by constructing a sequence of terms as follows. Choose terms $p_1, p_2, \ldots$ up to the first index k_1 such that

$$s < p_1 + p_2 + \cdots + p_{k_1}.$$

4D*. Euler's Constant and Summation

This will eventually occur, because $\sum_1^\infty p_n = \infty$. Second, choose $q_1, q_2, \ldots$ up to the first index l_1 such that

$$(p_1 + p_2 + \cdots + p_{k_1}) + (q_1 + q_2 + \cdots q_{l_1}) < s.$$

Third, choose just enough new p's to get to the right of s, and then choose just enough new q's to get to the left of s, and continue. At each phase of the $2n$-th step of this construction, the difference between s and the partial sum of the new series has absolute value smaller than $\max\{p_{k_n}, |q_{l_n}|\}$, and at each phase of the $2n+1$-st step, the difference has absolute value smaller than $\max\{p_{k_{n+1}}, |q_{l_n}|\}$. These have limit 0, so the rearranged series has sum s. □

Exercise

1. Suppose that the real series $\sum_1^\infty a_n$ is conditionally convergent. Show that it has a rearrangement that diverges to $+\infty$.

4D*. Euler's Constant and Summation

Let us take another look at the harmonic series $\sum_1^\infty 1/n$. As in (9), we can see that its partial sums are approximated by integrals. Here we look at

$$\sum_{k=1}^n \frac{1}{k} > \sum_{k=1}^n \int_k^{k+1} \frac{dx}{x} = \int_1^{n+1} \frac{dx}{x} = \log(n+1),$$

where, again, $\log(n+1)$ denotes the natural logarithm. Let us look at the difference

$$\gamma_n = \sum_{k=1}^n \frac{1}{k} - \log(n+1) = \sum_{k=1}^n \int_k^{k+1} \left(\frac{1}{k} - \frac{1}{x} \right) dx$$

$$= \sum_{k=1}^n \int_k^{k+1} \frac{x-k}{xk} dx. \tag{16}$$

This identity makes it clear that the sequence $\{\gamma_n\}$ is positive and strictly increasing. Moreover, the integrand of the $k+1$-th summand is $\leq 1/x(k+1) \leq 1/(x-1)^2$, so the sequence is bounded with limit

$$0 < \gamma < 1 + \int_2^\infty \frac{dx}{(x-1)^2} = 2.$$

The constant γ is called *Euler's constant* and has value $.5772\ldots$. Not much is known about γ – even whether it is rational – but its very existence allows us to obtain exact evaluations of the series (10) and (11). In fact, let $\{s_n\}$ be the partial sums of the harmonic series. According to (16), $s_n = \log(n+1) + \gamma_n$. Therefore

$$1 - \frac{1}{2} + \frac{1}{3} - \frac{1}{4} + \cdots + \frac{1}{2n-1} - \frac{1}{2n}$$
$$= \left(1 + \frac{1}{2} + \frac{1}{3} + \cdots + \frac{1}{2n}\right) - 2\left(\frac{1}{2} + \frac{1}{4} + \frac{1}{6} + \cdots + \frac{1}{2n}\right)$$
$$= s_{2n} - 2(s_n/2) = s_{2n} - s_n = (\log(2n+1) + \gamma_{2n}) - (\log(n+1) + \gamma_n)$$
$$= \log \frac{2n+1}{n+1} + (\gamma_{2n} - \gamma_n).$$

Because $\{\gamma_n\}$ converges, it follows that the limit of the preceding is $\log 2$, that is,

$$1 - \frac{1}{2} + \frac{1}{3} - \frac{1}{4} + \cdots + \frac{1}{2n-1} - \frac{1}{2n} + \cdots = \log 2,$$

as promised in Chapter 1.

We leave as an exercise the similar, slightly trickier, argument that shows that the series (11) has sum $(3 \log 2)/2$.

Exercise

1. Prove that the sum of the series (11) is $(3 \log 2)/2$.

4E*. Conditional Convergence: Summation by Parts

The following generalizes the theorem on alternating series. The idea is due to Abel.

Theorem 4.13. *Suppose that the sequence $\{a_n\}$ has the properties*

$$a_1 \geq a_2 \geq a_3 \geq \cdots \geq a_n \geq 0; \qquad \lim_{n \to \infty} a_n = 0, \tag{17}$$

and suppose that the sequence $\{B_n\}$ of partial sums of the series $\{b_n\}$ is bounded. Then the series $\sum_1^\infty a_n b_n$ converges.

Before getting to the proof we consider two examples. First, for alternating series $\sum_1^\infty (-1)^{n-1} a_n$ we take $b_n = (-1)^{n-1}$. Then the partial sums are $B_{2n-1} = 1$ and $B_{2n} = 0$, so Theorem 4.13 implies Theorem 4.10. Next, consider the series

$$z + \frac{z^2}{2} + \frac{z^3}{3} + \frac{z^4}{4} + \cdots, \tag{18}$$

where z is some complex number. The ratio $|a_{n+1}|/|a_n|$ for this series is $|z|(1 + 1/n)$, which has limit $|z|$, so the series converges absolutely if $|z| < 1$ and the terms do not have limit 0 if $|z| > 1$. Therefore the interesting question is: What happens if

$|z| = 1$? For $z = 1$, (18) is the (divergent) harmonic series and for $z = -1$ it is the (convergent) alternating harmonic series. For other z's having modulus 1 we try Theorem 4.13 with $a_n = 1/n$ and $b_n = z^n$. Then

$$|B_n| = \left|\sum_{k=1}^{n} z^k\right| = \frac{|z - z^{n+1}|}{|1 - z|} \leq \frac{|z| + |z|^{n+1}}{|1 - z|} = \frac{2}{|1 - z|}.$$

Thus (18) converges if and only if $|z| < 1$ (absolute convergence) or $|z| = 1$ but $z \neq 1$ (conditional convergence).

Proof of Theorem 4.13: Note that $b_n = B_n - B_{n-1}$. If $m < n$, the difference between the n-th and m-th partial sums of $\sum_1^\infty a_n b_n$ is

$$a_{m+1}b_{m+1} + a_{m+2}b_{m+2} + \ldots + a_{n-1}b_{n-1} + a_n b_n$$
$$= a_{m+1}(B_{m+1} - B_m) + a_{m+2}(B_{m+2} - B_{m+1}) + \ldots + a_n(B_n - B_{n-1})$$
$$= -a_{m+1}B_m + (a_{m+1} - a_{m+2})B_{m+1} + \ldots + (a_{n-1} - a_n)B_{n-1} + a_n B_n. \quad (19)$$

Suppose that $|B_k| \leq K$, all n. Note that the a_k's and the $a_k - a_{k+1}$'s are nonnegative. Take the modulus of each side in (19) to see that the modulus of this difference between partial sums is at most

$$[a_{m+1} + (a_{m+1} - a_{m+2}) + \cdots + (a_{n-1} - a_n) + a_n] K = 2a_{m+1}K.$$

But $\{a_m\}$ has limit 0, so the sequence of partial sums of $\sum_1^\infty a_n b_n$ is a Cauchy sequence. □

The computation (19), which converted the sum of terms $a_k b_k$ into a sum of terms involving *sums* of the b_k's and *differences* of the a_k's, is called *summation by parts*. It is closely analogous to integration by parts, which converts the integral of a product fg into the integral of $f'G$, where G is an *integral* of g and f' is the *derivative* of f.

Additional Exercises for Chapter 4

1. Suppose that b is positive and $a_n = 1/n^b$. Compute the limit

$$\lim_{n \to \infty} n \left(\frac{a_n}{a_{n+1}} - 1\right).$$

2. (Raabe's test) Suppose that $a_n > 0$, all n. Prove that $\sum_1^\infty a_n$ converges if

$$\liminf_{n \to \infty} n \left(\frac{a_n}{a_{n+1}} - 1\right) > 1$$

and diverges if $n(a_n/a_{n+1} - 1) \leq 1$, all $n \geq N$.

3. (Gauss's test) Suppose that $a_n > 0$, all n. Suppose that $p > 1$ and that
$$n^p \left| \frac{a_n}{a_{n+1}} - 1 - \frac{\alpha}{n} \right|$$
is bounded. Prove that $\sum_1^\infty a_n$ converges if $\alpha > 1$ and diverges if $\alpha \leq 1$.

4. (Bertrand's test) Suppose that $a_n > 0$, all n. Prove that $\sum_1^\infty a_n$ converges if
$$\liminf_{n \to \infty} \log n \left\{ n \left(\frac{a_n}{a_{n+1}} - 1 \right) - 1 \right\} > 1$$
and diverges if
$$\limsup_{n \to \infty} \log n \left\{ n \left(\frac{a_n}{a_{n+1}} - 1 \right) - 1 \right\} < 1.$$

5
Power Series

We shall use power series to define some of the most important functions of analysis. To begin, we establish the basic properties: radius of convergence, differentiation, and products.

5A. Power Series, Radius of Convergence

Suppose that $\{a_n\}_0^\infty$ is a complex sequence indexed by the nonnegative integers. The associated *power series* is (are) the series

$$\sum_{n=0}^\infty a_n z^n = a_0 + a_1 z + a_2 z^2 + a_3 z^3 + \cdots, \qquad z \in \mathbb{C}. \tag{1}$$

(The standard convention is that $z^0 = 1$ even for $z = 0$.) An example is the geometric series $\sum_0^\infty z^n$. An application of the Ratio Test shows that the geometric series converges if $|z| < 1$ and diverges if $|z| > 1$. This dichotomy is characteristic of power series.

Theorem 5.1: Radius of convergence. *Suppose that $\{a_n\}_0^\infty$ is a complex sequence. There is an extended real R, $0 \leq R \leq \infty$, such that the series (1) converges absolutely when $|z| < R$ and diverges when $|z| > R$. The number R, called the radius of convergence of the series, is*

$$R = \frac{1}{\limsup_{n \to \infty} |a_n|^{1/n}}. \tag{2}$$

In particular, $R = 0$ means that (1) converges only for $z = 0$ and $R = \infty$ means that (1) converges for every complex z.

Proof: This is a simple application of the Root Test, taking into account the presence of the powers z^n. In fact,

$$\limsup_{n\to\infty} |a_n z^n|^{1/n} = \limsup_{n\to\infty} |a_n|^{1/n}|z| = \left(\limsup_{n\to\infty} |a_n|^{1/n}\right)|z|.$$

This upper limit is less than 1 if $|z| < R$ and greater than 1 if $|z| > R$. □

Remarks. 1. Theorem 5.1 says nothing about convergence when $|z| = R$, if R is positive and finite. The reason is that it depends on the particular series. Consider three examples:

$$\sum_{n=0}^{\infty} z^n, \quad \sum_{n=1}^{\infty} \frac{z^n}{n}, \quad \sum_{n=1}^{\infty} \frac{z^n}{n^2}.$$

(For the last two, the coefficient of z^0 is 0.) Suppose that $|z| = 1$. The first diverges, since the terms have modulus 1. The third converges absolutely, by comparison to $\sum_{1}^{\infty} 1/n^2$. The second diverges when $z = 1$ (harmonic series) and converges when $z = -1$ (alternating harmonic series). Indeed, the second series converges conditionally for all z such that $|z| = 1$ but $z \neq 1$ (see Section 4E).

2. Although (2) gives a definitive formula for R, it is frequently hard to compute. The following theorem is often useful.

Theorem 5.2. *Suppose that $\{a_n\}_0^\infty$ is a complex sequence with the properties: $a_n \neq 0$ for $n \geq N$ and*

$$\lim_{n\to\infty} \frac{|a_{n+1}|}{|a_n|}$$

exists (possibly $= \infty$). Then the radius of convergence of $\sum_{n=0}^{\infty} a_n z^n$ is the reciprocal

$$R = \lim_{n\to\infty} \frac{|a_n|}{|a_{n+1}|}. \tag{3}$$

Proof: This is an application of the Ratio Test and is left as an exercise.

Examples. The series $\sum_{0}^{\infty} n!\, z^n$ has radius of convergence $R = 0$. The series $\sum_{0}^{\infty} z^n/n!$ has radius of convergence $R = \infty$, as do the series

$$\sum_{n=0}^{\infty} (-1)^n \frac{z^{2n}}{(2n)!} = 1 - \frac{z^2}{2} + \frac{z^4}{24} - \frac{z^6}{720} + \cdots;$$

$$\sum_{n=0}^{\infty} (-1)^n \frac{z^{2n+1}}{(2n+1)!} = z - \frac{z^3}{6} + \frac{z^5}{120} - \frac{z^7}{5040} + \cdots.$$

[We shall have much to say about these series in Chapter 9.]

Exercises

In Exercises 1–8, determine the radius of convergence.

1. $$\sum_{n=0}^{\infty} n^a z^n, \quad a \text{ real.}$$

2. $$\sum_{n=1}^{\infty} 3^n z^n / n^3.$$

3. $$\sum_{n=0}^{\infty} a^{n^2} z^n, \quad a > 0.$$

4. $$\sum_{n=0}^{\infty} \frac{(n!)^2}{(2n)!} z^n.$$

5. $$\sum_{n=0}^{\infty} n! \, z^{n!}.$$

6. $$\sum_{n=1}^{\infty} \frac{n!}{n^n} z^n.$$

7. $$\sum_{n=0}^{\infty} a(a+1)(a+2)\cdots(a+n-1) z^n, \quad a > 0.$$

8. $$\sum_{n=0}^{\infty} \frac{a(a+1)(a+2)\cdots(a+n-1)}{b(b+1)(b+2)\cdots(b+n-1)} z^n, \quad b \notin \mathbb{Z}.$$

9. Prove Theorem 5.2
10. Suppose that $\sum_0^\infty a_n z^n$ has radius of convergence R and $0 < r < R$.
    (a) Prove that there is a constant K such that $|z| \leq r$ implies $|\sum_0^\infty a_n z^n| \leq K$. (In fact, one can take $K = \sum_0^\infty |a_n| r^n$, but why is this finite?)
    (b) Prove that for any given $k \in \mathbb{N}$ there is a constant K_k such that
    $$\left| \sum_{n=k}^{\infty} a_n z^n \right| \leq K_k |z|^k \quad \text{if } |z| \leq r.$$

5B. Differentiation of Power Series

Suppose that $\sum_0^\infty a_n z^n$ is a power series with radius of convergence $R > 0$. Then it defines a *function* of z in the disk of radius R centered at the origin (the *disk of convergence*):

$$f(z) = \sum_{n=0}^{\infty} a_n z^n = a_0 + a_1 z + a_2 z^2 + a_3 z^3 + \cdots, \quad |z| < R. \quad (4)$$

We show here that this series may be differentiated term by term. The derivative of a function of a complex variable is defined in the same way as for a function of a real variable, as the limit of difference quotients. Suppose that f is defined for

$|z| < R$. Given a point z with $|z| < R$, the complex number c is the *derivative* of f at z, written $c = f'(z)$, if

$$\lim_{w \to z} \frac{f(w) - f(z)}{w - z} = c.$$

The more precise version of this is, for each $\varepsilon > 0$ there is a $\delta > 0$ such that, whenever $0 < |z - w| < \delta$ and $|w| < R$, it follows that

$$\left| \frac{f(w) - f(z)}{w - z} - c \right| < \varepsilon. \tag{5}$$

If the usual rules of calculus applied even for infinite sums, we could expect that the derivative of (4) would be the power series

$$\sum_{n=0}^{\infty} (n+1) a_{n+1} z^n = \sum_{n=1}^{\infty} n a_n z^{n-1} = a_1 + 2a_2 z + 3a_3 z^2 + \cdots. \tag{6}$$

In fact this is correct. To prove it we first show that (6) and *its* formal derivative both converge.

Lemma 5.3. *If the series* (1) *has radius of convergence* R, *then the series*

$$\sum_{n=0}^{\infty} (n+1) a_{n+1} z^n, \qquad \sum_{n=0}^{\infty} (n+2)(n+1) a_{n+2} z^n \tag{7}$$

also have radius of convergence R.

Proof: The second series in (7) has the same relation to the first as the first does to the series (6), so we only need to consider the first. Multiplication by z gives the series $\sum_0^{\infty} n a_n z^n$, which therefore converges for the same values of z and has the same radius of convergence as $\sum_0^{\infty} (n+1) a_{n+1} z^n$ itself. By Proposition 3.2,

$$\limsup_{n \to \infty} |n a_n|^{1/n} = \left(\lim_{n \to \infty} n^{1/n} \right) \left(\limsup_{n \to \infty} |a_n|^{1/n} \right) = 1 \cdot \limsup_{n \to \infty} |a_n|^{1/n},$$

so this series has radius of convergence R. □

Theorem 5.4: Differentiation of power series. *If the power series* $\sum_0^{\infty} a_n z^n$ *has radius of convergence* $R > 0$, *then the associated function* $f(z) = \sum_0^{\infty} a_n z^n$ *is differentiable at each point of the disk of convergence and*

$$f'(z) = \sum_{n=1}^{\infty} n a_n z^{n-1} = a_1 + 2a_2 z + 3a_3 z^2 + \cdots, \qquad |z| < R. \tag{8}$$

5B. Differentiation of Power Series

Proof: Given z with modulus $|z| < R$, we fix a number r, $|z| < r < R$. Suppose that $|w - z| < r - |z|$. This implies that $|w| < r$. We want to show that the modulus of

$$\frac{f(w) - f(z)}{w - z} - \sum_{n=1}^{\infty} n a_n z^{n-1} \qquad (9)$$

is small when $|z - w|$ is small. Now f is defined by (4), and we may use the algebraic properties (Theorem 4.2) to write (9) as the sum of a series whose terms are

$$a_n \left(\frac{w^n - z^n}{w - z} - n z^{n-1} \right) = a_n (w^{n-1} + w^{n-2} z + \cdots + z^{n-1} - n z^{n-1})$$

$$= a_n [(w^{n-1} - z^{n-1}) + (w^{n-2} - z^{n-2})z + \cdots + (w - z) z^{n-2}] \qquad (10)$$

$$= a_n (w - z) \left[\frac{w^{n-1} - z^{n-1}}{w - z} + \frac{w^{n-2} - z^{n-2}}{w - z} z + \cdots + \frac{w - z}{w - z} z^{n-2} \right].$$

Since $|z| < r$ and $|w| < r$,

$$\left| \frac{w^k - z^k}{w - z} \right| = |w^{k-1} + w^{k-2} z + \cdots + z^{k-1}| < k r^{k-1}. \qquad (11)$$

It follows from (11) that (10) has modulus

$$< |w - z| |a_n| [(n - 1) r^{n-2} + (n - 2) r^{n-3} r + \cdots + 2r r^{n-3} + r^{n-2}]$$

$$= |w - z| |a_n| r^{n-2} [(n - 1) + (n - 2) + \cdots + 1]$$

$$= \tfrac{1}{2} n(n - 1) |a_n| |w - z| r^{n-2}. \qquad (12)$$

The power series with coefficients $|a_n|$ has the same radius of convergence as that with coefficients a_n. Therefore we deduce from Lemma 5.3 that

$$\sum_{n=2}^{\infty} n(n - 1) |a_n| r^{n-2} = K < +\infty. \qquad (13)$$

Putting all this together we obtain

$$\left| \frac{f(w) - f(z)}{w - z} - \sum_{n=1}^{\infty} n a_n z^{n-1} \right| < \tfrac{1}{2} K |w - z| \qquad \text{if } |w - z| < r - |z|, \qquad (14)$$

and it follows that the derivative is given by (8). □

Consider the converse situation: Suppose that we know that a function like $f(z) = 1/(1 - z)$ is given by a power series; how can we determine the coefficients of the series from the function itself? It is clear from (1) that $f(0) = a_0$. Differentiating gives $f'(0) = a_1$, $f''(0) = 2a_2$, $f'''(0) = 3 \cdot 2a_3$, and so on.

Corollary 5.5: Coefficients and derivatives. *If the power series $\sum_0^\infty a_n z^n$ has a positive radius of convergence and define the function f, then*

$$a_k = \frac{f^{(k)}(0)}{k!}, \qquad k = 0, 1, 2, \ldots. \tag{15}$$

Exercises

1. For $|w - 1| < 1$, let $f(w) = \sum_{n=1}^\infty (w - 1)^n/n$. Find $f'(w)$.
2. Determine the coefficients $\{a_n\}_0^\infty$ of the power series whose sum is $(1 - z)^{-2}$.
3. Determine the sum of the series $\sum_0^\infty (n + 2)(n + 1)z^n$, $|z| < 1$.
4. Prove that, for any positive integer k,

$$\sum_{n=0}^\infty \binom{k+n}{k} z^n = \frac{1}{(1-z)^{k+1}} \qquad |z| < 1.$$

5. Suppose that $\sum_0^\infty a_n z^n$ has radius of convergence $R > 0$, and suppose that $|z_0| = r < R$. Define

$$g(z) = \sum_{n=0}^\infty a_n (z - z_0)^n, \qquad |z - z_0| < R - r.$$

Prove that g is given by a convergent power series

$$g(z) = \sum_{n=0}^\infty b_n z^n$$

whose radius of convergence is at least $R - r$.
6. Suppose that the function f is defined by a convergent power series and suppose that $f(z + w) = f(z)f(w)$ for all complex z, w.
   (a) Prove directly from this assumption that there is a constant $a \in \mathbb{C}$ such that $f'(z) = af(z)$, all z. (In fact, $a = f'(0)$.)
   (b) Use (a) to prove that $f(z) = \sum_0^\infty (az)^n/n!$.
7. Determine the coefficients of the power series that defines a function with the following properties: $f''(z) = -f(z)$, $f(0) = 1$, $f'(0) = 0$.

5C. Products and the Exponential Function

We begin by defining the product of two series and then apply the result to products of power series. Consider the formal product

$$(a_0 + a_2 + a_2 + \cdots)(b_0 + b_2 + b_2 + \cdots). \tag{16}$$

5C. Products and the Exponential Function

Each term $a_j b_k$ occurs in the formal product. To be able to consider this as a series we need to group these terms so as to sum over a single index. One way is to group the $a_j b_k$'s whose indices have the same sum. The result is the formal series

$$a_0 b_0 + (a_0 b_1 + a_1 b_0) + (a_0 b_2 + a_1 b_1 + a_2 b_0) + \cdots . \tag{17}$$

This series $\sum_0^\infty c_n$ with terms

$$c_n = a_0 b_n + a_1 b_{n-1} + \cdots + a_{n-1} b_1 + a_n b_0 \tag{18}$$

is called the *Cauchy Product* or simply the *product* of the series $\sum_0^\infty a_n$ and $\sum_0^\infty b_n$.

One might expect that the product of convergent series is convergent, with its sum equal to the product of the two sums. Now the alternating series

$$\sum_{n=0}^\infty \frac{(-1)^n}{\sqrt{n+1}} = 1 - \frac{1}{\sqrt{2}} + \frac{1}{\sqrt{3}} - \frac{1}{\sqrt{4}} + \cdots$$

converges, but the term c_{n-1} in the product of this series with itself is

$$c_{n-1} = (-1)^{n-1} \left(\frac{1}{\sqrt{n}} + \frac{1}{\sqrt{2}\sqrt{n-1}} + \cdots + \frac{1}{\sqrt{n}} \right).$$

There are n summands in parentheses, each $\geq (1/\sqrt{n})^2$, so $|c_{n-1}| \geq 1$ and the product diverges.

This discouraging situation cannot arise if one of the factors is absolutely convergent, by a theorem of Mertens.

Theorem 5.6: Products of series. *If the series $\sum_0^\infty a_n$ is absolutely convergent with sum A and the series $\sum_0^\infty b_n$ converges with sum B, then the product series $\sum_0^\infty c_n$ converges, and its sum is AB.*

Proof: Let $\{A_n\}$, $\{B_n\}$, and $\{C_n\}$ be the sequences of partial sums. Since $A_n B_n$ has limit AB, it is enough to show that $A_n B_n - C_n$ has limit 0. Suppose that $m < n$. Then

$$A_n B_n - C_n = \sum_{j,k \leq n;\ j+k > n} a_j b_k = \sum_{j=1}^n a_j \left(\sum_{k=n-j+1}^n b_k \right)$$

$$= \sum_{j=0}^m a_j (B_n - B_{n-j}) + \sum_{j=m+1}^n a_j (B_n - B_{n-j}). \tag{19}$$

Let K be a bound for the $|B_n|$ and let $L = \sum_0^\infty |a_n|$. Given $\varepsilon > 0$, choose m so large and then n so large that

$$\sum_{j=m+1}^\infty |a_j| < \varepsilon; \qquad |B_n - B_{n-j}| < \varepsilon \qquad \text{if } 0 \le j \le m.$$

Because of these choices, the first sum in the last line of (19) has modulus $\le L\varepsilon$ and the second sum has modulus $\le 2K\varepsilon$. Therefore $A_n b_n - C_n$ has limit 0. □

We turn now to the product of power series.

Theorem 5.7: Multiplication of power series. *The product of the power series* $\sum_0^\infty a_n z^n$ *and* $\sum_0^\infty b_n z^n$ *is the power series* $\sum_0^\infty c_n z^n$,

$$c_n = a_0 b_n + a_1 b_{n-1} + \cdots + a_{n-1} b_1 + a_n b_0.$$

The radius of convergence of the product is at least as large as the smaller of the radii of convergence of the factors.

Proof: The first statement follows from the definition of the product: In formula (18), the term $a_j b_{n-j}$ is to be replaced by

$$(a_j z^j)(b_{n-j} z^{n-j}) = a_j b_{n-j} z^n.$$

The second statement follows from Theorem 5.6. □

Consider the apparently unrelated matter of powers of a positive number. The positive integer powers of $a > 0$ satisfy the equations $a^{m+n} = a^m a^n$. This equation carries over to *all* integral powers if and only if one defines $a^0 = 1$, $a^{-n} = 1/a^n$, $n \in \mathbb{N}$. The equation carries over to all rational powers if and only if one defines $a^{m/n} = (a^{1/n})^m$, $m \in \mathbb{Z}$, $n \in \mathbb{N}$. We now try to extend this to real powers; if $f(z) = a^x$ has been defined, we want $f(x+y) = f(x)f(y)$, for all real x and y.

Let us be more ambitious and look for functions f, defined for all *complex* z, that satisfy

$$f(z+w) = f(z)f(w), \qquad \text{all complex } z, w. \tag{20}$$

In fact, we shall find *every* such function that is defined by a convergent power series, by determining what the coefficients can be. Suppose that $f(z) = \sum_0^\infty a_n z^n$. Fix z and w for the moment and consider $f(tz + tw)$ for complex t. This is a power

5C. Products and the Exponential Function

series in t and (20) is equivalent to

$$\sum_{n=0}^{\infty} a_n(z+w)^n t^n = \sum_{0}^{\infty} \left(\sum_{j=0}^{n} a_j a_{n-j} z^j w^{n-j} \right) t^n, \qquad t, z, w \in \mathbb{C}. \qquad (21)$$

Corollary 5.5 implies that the power series in t in (21) are equal if and only if the coefficients are the same:

$$a_n(z+w)^n = \sum_{j=0}^{n} a_j a_{n-j} z^j w^{n-j}, \qquad n = 0, 1, 2, \ldots. \qquad (22)$$

This is true for all z and w if and only if the coefficients of each $z^j w^{n-j}$ are the same after the left-hand side is expanded. This means

$$j! a_j \cdot (n-j)! a_{n-j} = n! a_n, \qquad \text{all } n, \text{ all } j \leq n. \qquad (23)$$

Set $b_k = k! a_k$. The equations (23) are equivalent to the equations $b_m b_n = b_{n+m}$. Let $b_1 = a_1 = a$ be given. Then the unique solution is $b_m = a^m$, $m \geq 0$. Thus $a_m = a^m / m!$. What we have shown is that a function f given by a convergent power series satisfies (20) *if and only* if it has the form

$$\sum_{n=0}^{\infty} \frac{a^n z^n}{n!} = 1 + az + \frac{(az)^2}{2} + \frac{(az)^3}{6} + \cdots$$
$$= E(az), \qquad \text{for some complex } a,$$

where

$$E(z) = \sum_{n=0}^{\infty} \frac{z^n}{n!} = 1 + z + \frac{z^2}{2} + \frac{z^3}{6} + \frac{z^4}{24} + \frac{z^5}{120} + \cdots. \qquad (24)$$

This series converges for every $z \in \mathbb{C}$.

Theorem 5.8: The exponential function. *The function E defined by* (24) *has the properties*

$$E(z+w) = E(z) E(w), \quad \text{all } z, w \in \mathbb{C}; \qquad E(r) = e^r \quad \text{all } r \in \mathbb{Q}. \qquad (25)$$

Here e is the positive real number

$$e = E(1) = 1 + 1 + \frac{1}{2} + \frac{1}{6} + \frac{1}{24} + \frac{1}{120} + \cdots + \frac{1}{n!} + \cdots. \qquad (26)$$

Proof: We showed above that the functions $f(z) = E(az)$ are the only solutions of (20) given by power series; in particular, E itself satisfies (20). Note that $x > 0$

implies $E(x) > 0$. Repeated application of (20) gives, for $m, n \in \mathbb{N}$,

$$[E(1/n)]^n = E(1/n)E(1/n)\ldots E(1/n) = E(1/n + 1/n + \ldots + 1/n)$$
$$= E(1) = e;$$
$$[E(1/n)]^m = E(1/n)E(1/n)\ldots E(1/n) = E(1/n + 1/n + \ldots + 1/n)$$
$$= E(m/n).$$

The first of these equations tells us that $E(1/n) = e^{1/n}$ and the second then gives $E(m/n) = e^{m/n}$. Now $E(0) = 1 = e^0$ and $E(-m/n)E(m/n) = E(0) = 1$, so $E(-m/n) = 1/E(m/n) = e^{-m/n}$. □

Exercises

1. Determine the coefficients $\{a_n\}_0^\infty$ of the power series whose sum is $(1-z)^{-2}$ for $|z| < 1$, by squaring $(1-z)^{-1}$.
2. Determine the coefficients of the power series that defines a function with the following properties: $f''(z) = 2f'(z) - f(z)$, $f(0) = 0$, $f'(0) = 1$. How is this function related to the exponential function?
3. Prove that e is irrational. (Show that if $\{s_n\}_0^\infty$ is the sequence of partial sums of $\sum_0^\infty 1/n!$, then

$$0 < e - s_n < \frac{1}{(n+1)!}\left(1 + \frac{1}{n+1} + \frac{1}{(n+1)^2} + \cdots\right) = \frac{1}{n!\, n}.$$

If e were the rational p/q, then $q!\, e$ and $q!\, s_q$ would be [distinct] integers.)
4. (Abel's product theorem) Suppose that the series $\sum_0^\infty a_n$ and $\sum_0^\infty b_n$ converge and have sums A and b, respectively. Suppose also that the Cauchy Product $\sum_0^\infty c_n$ converges, with sum C. Prove that $C = AB$. (None of these series is assumed to be absolutely convergent.)

5D*. Abel's Theorem and Summation

Here we look a second time at the problem of summing the alternating harmonic series $\sum_1^\infty (-1)^{n-1}/n$. This time we consider the associated power series

$$f(z) = \sum_{n=1}^\infty (-1)^{n-1}\frac{z^n}{n} = z - \frac{z^2}{2} + \frac{z^3}{3} - \frac{z^4}{4} + \cdots. \qquad (27)$$

The radius of convergence is 1 and the derivative is the geometric series

$$f'(z) = 1 - z + z^2 - z^3 + \ldots = \sum_{n=0}^\infty (-z)^n = \frac{1}{1+z}.$$

5D*. Abel's Theorem and Summation

Note that $f(0) = 0$. Thus we may integrate to obtain

$$f(x) = \int_0^x \frac{dt}{1+t} = \log(1+x), \qquad 0 \le x < 1. \tag{28}$$

It is tempting now to take the limit on both sides and conclude that $\sum_1^\infty (-1)^{n-1}/n = \log 2$. But can one justify this procedure?

Theorem 5.9: Abel's Summation Theorem. *Suppose that the power series $\sum_0^\infty a_n z^n$ has radius of convergence 1 and suppose that the series $\sum_0^\infty a_n$ converges. Then*

$$\lim_{x \to 1,\, x<1} \sum_{n=0}^\infty a_n x^n = \sum_{n=0}^\infty a_n. \tag{29}$$

Proof: Let $\{B_n\}$ be the sequence with terms

$$B_n = \sum_{k=n}^\infty a_k = a_n + a_{n+1} + a_{n+2} + \cdots. \tag{30}$$

Note that $a_n = B_n - B_{n+1}$. Given $\varepsilon > 0$, choose m so large that $n \ge m$ implies $|B_n| < \varepsilon$ and fix m. For $n > m$,

$$\sum_{k=0}^n a_k - \sum_{k=0}^n a_k x^k = \sum_{k=0}^m a_k(1-x^k) + \sum_{k=m+1}^n a_k(1-x^k). \tag{31}$$

Since m is fixed, the first of the two sums on the right has limit 0 as $x \to 1$. The proof will be complete if we show that for each $n > m$ the modulus of the second of these sums is $< 2\varepsilon$. To do so we set $y_k = 1 - x^k$ and note that

$$0 < y_k \le 1; \qquad 0 \le x^k - x^{k+1} = y_{k+1} - y_k < 1, \quad \text{if } 0 \le x < 1. \tag{32}$$

We reorganize the second sum:

$$a_{m+1} y_{m+1} + a_{m+2} y_{m+2} + \cdots + a_n y_n$$
$$= (B_{m+1} - B_{m+2}) y_{m+1} + \cdots + (B_n - B_{n+1}) y_n$$
$$= B_{m+1} y_{m+1} + B_{m+2}(y_{m+2} - y_{m+1}) + \cdots + B_n(y_n - y_{n-1}) - B_{n+1} y_n. \tag{33}$$

Because of (32) and the fact that $|B_k| < \varepsilon$ for $k > m$, the modulus of the sum (33) is at most

$$\varepsilon [y_{m+1} + (y_{m+2} - y_{m+1}) + \cdots + (y_n - y_{n-1}) + y_n] = 2\varepsilon\, y_n \le 2\varepsilon. \qquad \square$$

The reader is invited to critique the following argument. Differentiating the geometric series $\sum z^n = 1/(1-z)$ gives

$$\frac{1}{(1-z)^2} = \sum_{n=1}^{\infty} n z^{n-1}.$$

Taking $z = -1$, we get

$$\frac{1}{4} = \sum_{n=1}^{\infty} (-1)^{n-1} n = 1 - 2 + 3 - 4 + 5 - 6 + \cdots.$$

This allows us to evaluate

$$S = 1 + 2 + 3 + 4 + 5 + 6 + \cdots.$$

In fact,

$$\begin{aligned}
\tfrac{1}{4} &= 1 - 2 + 3 - 4 + 5 - 6 + \cdots \\
&= (1 + 2 + 3 + 4 + 5 + 6 + \cdots) - 2(2 + 4 + 6 + 8 + \cdots) \\
&= (1 + 2 + 3 + 4 + 5 + 6 + \cdots) - 4(1 + 2 + 3 + 4 + 5 + 6 + \cdots) \\
&= -3S.
\end{aligned}$$

Thus $S = -1/12$, as promised in Chapter 1A.

Exercises

1. Use the method of Section 5D to obtain the evaluation
$$1 + \frac{1}{3} - \frac{1}{2} + \frac{1}{5} + \frac{1}{7} - \frac{1}{4} + \cdots = \frac{3}{2} \log 2.$$

2. Prove the following partial converse of Abel's Theorem, due to Tauber: Suppose that $\sum_0^{\infty} a_n z^n$ has radius of convergence 1. Let $f(x) = \sum_0^{\infty} a_n z^n$ for $0 \leq x < 1$ and suppose that:

$$\lim_{0 \leq x < 1,\ x \to 1} f(x) = A; \qquad \lim_{n \to \infty} n a_n = 0.$$

Prove that $\sum_0^{\infty} a_n = A$. (This was the first "Tauberian theorem." Hardy and Littlewood showed that the result is still true if the second condition is weakened to $|a_n| \leq K/n$.)

6
Metric Spaces

Some of the concepts with which we have been concerned are not special to the real or complex numbers but are connected to the general idea of *distance*, in particular the notion of objects being very close to each other. These ideas can be made general and precise. The abstract concept allows us to develop once and for all a vocabulary and basic results that apply in many circumstances beyond the real numbers. In particular, there are various useful senses in which *functions* might be considered as being very close to each other (or not).

6A. Metrics

The abstract setting for the notion of distance is a set S, whose elements may be referred to as *points*. A distance function or *metric* on S is a function that assigns to each pair of points p and q in S a real number $d(p, q)$ and has the properties, for all p, q, and r in S:

D1 (Positivity) $d(p, q) \geq 0$; $d(p, q) = 0$ if and only if $p = q$.
D2 (Symmetry) $d(p, q) = d(q, p)$.
D3 (Triangle inequality) $d(p, r) \leq d(p, q) + d(q, r)$.

Definition. A metric space is a pair (S, d), where d is a metric defined on the set S.

Examples

1. If S is an arbitrary set, the *discrete metric* on S is defined by $d(p, p) = 0$, while $d(p, q) = 1$ if $q \neq p$.
2. Suppose that S consists of all strings of length n of 0's and 1's; two such strings of length 5 are 00101 and 10110. Let $d(p, q)$ be the number of places in which the strings differ; for our example, $d(p, q) = 3$. (If q is supposed to be a copy of p, $d(p, q)$ is the number of errors.)

3. If (S, d) is a metric space and A is any subset of S, then A inherits the metric d and (A, d) is a metric space.
4. The *standard metric* in $\mathbb{C}$ and its subsets is $d(z, w) = |z - w|$. This gives the standard metric $d(x, y) = |x - y|$ in $\mathbb{R}$ and its subsets. If we identify $\mathbb{C}$ with the plane

$$\mathbb{R}^2 = \{\mathbf{x} = (x_1, x_2) : x_j \in \mathbb{R}\}$$

in the usual way, then the standard metric takes the form

$$d(\mathbf{x}, \mathbf{y}) = \sqrt{(y_1 - x_1)^2 + (y_2 - x_2)^2}.$$

5. Another example of a metric in $\mathbb{R}^2$ is

$$d_1(\mathbf{x}, \mathbf{y}) = |y_1 - x_1| + |y_2 - x_2|.$$

(If one thinks of this as the sum of the east/west and north/south distances, it is an appropriate notion of distance in a city that is laid out in a rectangular grid [and has no vacant lots].)

6. If V is a vector space over the real (or complex) numbers, a *norm* on V is a function from V to the nonnegative reals, $\mathbf{v} \to ||\mathbf{v}||$, that has the following properties for any vectors $\mathbf{v}$ and $\mathbf{w}$ in V and any scalar a in $\mathbb{R}$ (or $\mathbb{C}$):

(i) $\quad\quad\quad\quad ||\mathbf{v}|| \geq 0, \quad\quad ||\mathbf{v}|| = 0 \iff \mathbf{v} = \mathbf{0};$

(ii) $\quad\quad\quad\quad ||a\mathbf{v}|| = |a| \cdot ||\mathbf{v}||;$

(iii) $\quad\quad\quad\quad ||\mathbf{v} + \mathbf{w}|| \leq ||\mathbf{v}|| + ||\mathbf{w}||.$

A norm on V induces a metric on V: $d(\mathbf{v}, \mathbf{w}) = ||\mathbf{v} - \mathbf{w}||$ (as the reader should verify). Examples 4 and 5 are special cases.

Remarks. 1. It is common practice, when the metric is understood, to refer to the set S alone as a metric space. For example, when we consider $\mathbb{C}$ and its subsets as metric spaces, then we mean the standard metric *unless* we specify some other metric (e.g., the discrete metric). When we speak of an abstract set S as a metric space, S is considered as having some metric denoted d, or sometimes d_S.

2. It is usually easy to check properties (D1) and (D2) of a proposed metric. The triangle inequality may be more difficult, as in $\mathbb{C}$.

Exercises

1. Define two functions d_1 and d_∞ on $\mathbb{R}^n \times \mathbb{R}^n$ by

$$d_1(x, y) = \sum_{j=1}^{n} |x_j - y_j|, \quad\quad d_\infty(x, y) = \sup\{|x_j - y_j|, 1 \leq j \leq n\}.$$

(a) Show that these are both metrics on $\mathbb{R}^n$.

(b) Show that they are *equivalent* metrics in the sense that there is a positive constant C (which may depend on n) such that

$$C^{-1}d_1(x, y) \leq d_\infty(x, y) \leq Cd_1(x, y), \quad \text{all } x, y.$$

2. Given a metric d on S, define d^* by $d^*(p, q) = d(p, q)/[1 + d(p, q)]$. Prove that $d*$ is also a metric on S. Is it necessarily equivalent to d, in the sense of the preceding exercise?
3. Verify that a norm on a vector space induces a metric, as described above.

6B. Interior Points, Limit Points, Open and Closed Sets

Consider the following subsets of the complex plane:

$$A = \{z : |z| < 1\}; \qquad B = \{z : |z| \leq 1\}; \tag{1}$$
$$C = \{z : 0 < |z| \leq 1\}; \qquad D = \{z : |z| \leq 1;\ \mathrm{Re}\, z \in \mathbb{Q}\}.$$

These sets are qualitatively different. The concepts introduced in this section help to specify the differences in a precise way.

The basic concept is that of a *neighborhood* of a point in a metric space S. Given $p \in S$ and $r > 0$, the *r-neighborhood* of p is the set of points at distance $< r$ from p:

$$N_r(p) = \{q \in S : d(p, q) < r\}.$$

Thus the set A in (1) is $N_1(0)$ in $\mathbb{C}$.

Definition. A point p is said to be an *interior point* of a subset A of S if there is $\varepsilon > 0$ such that $N_\varepsilon(p) \subset A$. In other words, each point of S that is sufficiently close to p belongs to A. The set A is said to be *open* if each of its points is an interior point.

Note that if p is an interior point of A, then it is an interior point of any larger set B, $A \subset B \subset S$.

Examples. 1. The set A in (1) is open. The interior points of B are the points of A, the interior points of C are the nonzero points of A, and D has no interior points.

2. In any metric space S, the set S itself is open. The empty subset $\emptyset$ is also open — because for a set *not* to be open it must have some point that is not an interior point.

3. In $\mathbb{R}$, open intervals $(a, b) = \{x : a < x < b\}$ are open sets while the intervals $[a, b)$, $(a, b]$, $(a, b]$ are not.

4. No nonempty subset of $\mathbb{R}$ is open when considered as a subset of $\mathbb{C}$.

Remark. It can be helpful in working with these concepts to picture S as the plane and to make sketches. Sketches cannot substitute for proofs, but they can help to clarify concepts and to suggest proofs.

The following summarizes all but two of the basic general facts about open sets.

Proposition 6.1. *In a metric space S:*

(a) *Each neighborhood $N_r(p)$ of a point p in S is an open set.*
(b) *The intersection of a finite collection $A_1, A_2, \ldots, A_n$ of open sets is open.*
(c) *The union of any collection $\mathcal{A}$ of open sets is open.*

Proof: (a) Suppose that q belongs to $N_r(p)$. Let $s = r - d(p, q)$, which is positive by assumption. If q' is in $N_s(q)$, then the triangle inequality

$$d(p, q') \leq d(p, q) + d(q, q') < d(p, q) + [r - d(p, q)] = r$$

implies that q' is in $N_r(p)$. Thus q is an interior point of $N_r(p)$.

(b) If p is a point of the intersection, then there are positive numbers r_j such that $N_{r_j}(p) \subset A_j$. Let r be the smallest of the r_j's. Then $N_r(p)$ is included in the intersection.

(c) A point p that belongs to the union belongs to one of the open sets A and is therefore an interior point of the union, which is larger. □

Example. In connection with (b), consider the intersection of the sequence of open sets $A_n = N_{1+1/n}(0) \subset \mathbb{C}$. The intersection is the set B of (1), which is not open.

Definition. The *interior* of a subset B of S is the set whose elements are the interior points of B.

Proposition 6.2. *The interior of a set B is an open set and is the largest open set that is a subset of B.*

Proof: Let A be the interior of B. If p is in A, then a neighborhood $N_r(p)$ is a subset of B. By Proposition 6.1(a), each point of this neighborhood is an interior point of the neighborhood and therefore of B. Thus each point of the neighborhood is an interior point of B, so $N_r(p) \subset A$.

If $C \subset B$ is open, then each point of C is an interior point of B and so belongs to A. □

Definitions. A point p in S is said to be a *limit point* of a subset B of S if for each $\varepsilon > 0$ there is a point q in B, $q \neq p$, such that $q \in N_\varepsilon(p)$. An equivalent formulation is that *p is a limit point of B if it is the limit of a sequence of distinct* *points from B.* (See Section 6D for the general definition of limit. The proof of equivalence is left as an exercise.) A subset B of S is said to be *closed* if each limit point of B belongs to B.

Note that if p is a limit point of B, then p is a limit point of every larger set C, $B \subset C \subset S$.

Examples. 1. For each of the sets in (1), B is the set of limit points. Thus B is closed and the others are not. In particular, C and D are *neither open nor closed*.

2. In any metric space S, the set S itself is closed. The empty subset $\emptyset$ is also closed. Thus S and $\emptyset$ are *both open and closed*. [The reader should conclude from this and the preceding that common English usage is no substitute for careful thinking when handling these very precise concepts.]

3. In $\mathbb{R}$, closed intervals $[a, b] = \{x : a \leq x \leq b\}$ are closed sets while other types of bounded intervals are not.

4. In any metric space, any set that consists of a single point, or of finitely many points, is a closed set.

The following is the one remaining basic property of open sets; it is also the one general relationship between open sets and closed sets. Recall that the *complement* of a subset $A \subset S$ is the set $A^c = \{p \in S : p \notin A\}$.

Proposition 6.3. *A subset A of a metric space S is open if and only if its complement is closed.*

Proof: We begin with an observation about points of the complement that follows immediately from the definitions: *A point p in A is either a limit point of A^c or an interior point of A, but not both.* Thus, if A is open, then its points are not limit points of A^c, so limit points of A^c belong to A^c. If A^c is closed, then points of A are not limit points of A^c, so they are interior points of A and A is open. □

Remarks. The preceding proof illustrates two points. First, general results of this type are direct consequences of the definitions, which is a good reason to study the definitions. Second, it can be tempting to try indirect proofs ("suppose that A is open but A^c is not closed..."). One could give indirect proofs of both parts of the preceding proposition, but *when a direct proof is possible it is almost always shorter and more to the point*.

The following contains all but one of the remaining basic general properties of closed sets.

Proposition 6.4. *In any metric space S:*

(a) *The union of a finite collection B_1, B_2, ..., B_n of closed sets is closed.*
(b) *The intersection of any collection of closed sets is closed.*

Proof: (a) If p does not belong to the union of the B_j, then for each j there is $\varepsilon_j > 0$ such that $N_{\varepsilon_j}(p)$ is included in B_j^c. Let ε be the smallest of the ε_j. Then $N_\varepsilon(p) \subset B^c$, so p is not a limit point of the union. [What has happened here is that Proposition 6.1(b) and Proposition 6.3 have been combined in a somewhat disguised way. An alternative is to use them with no disguise.]

(b) If p is a limit point of the intersection, it is a limit point of each of the closed sets; therefore it belongs to each, and therefore it belongs to the intersection. □

Example. In connection with (a), consider the union of the sequence of closed intervals $[-1 + 1/n, 1 - 1/n]$ in $\mathbb{R}$.

Definition. The *closure* of a subset A of S is the set whose points are the points of A together with the limit points of A.

The following is the final basic property of closed sets.

Proposition 6.5. *The closure of a set A is closed and is the smallest closed set that includes A.*

Proof: Suppose that p is not in the closure of A. Then it has a neighborhood $N_r(p)$ that is included in A^c. This neighborhood is open, so none of its points is in the closure of A. Thus the complement of the closure is open, so the closure is closed.

If $C \supset A$ is closed, then each limit point of A is a limit point of C and therefore a point of C, so C includes the closure of A. □

Exercises

1. Suppose that S is a *finite* set with metric d. Prove that every subset of S is open.
2. Prove that p is a limit point of B if and only it is the limit of a sequence of distinct points from B.
3. Suppose that r is positive and p is a point of the metric space S. Prove that the subset $A = \{q \in S : d(p, q) \leq r\}$ is closed. Prove that $B = \{q \in S : d(p, q) > r\}$ is open.
4. Show that a point p is in the closure of a set A if and only if every neighborhood $N_\varepsilon(p)$ contains a point of A.
5. Prove that a subset of $\mathbb{R}$ has no interior points as a subset of $\mathbb{C}$.
6. Prove that any limit point in $\mathbb{C}$ of a subset of $\mathbb{R}$ must belong to $\mathbb{R}$.

7. Suppose that A is a closed subset of $\mathbb{R}$ that is bounded above. Prove that the least upper bound belongs to A.
8. Determine the closure of the set S in $\mathbb{R}^2$ that is the graph of the function $\sin(1/x)$, $S = \{(x, \sin(1/x)) : x \neq 0\}$.
9. Let int(A) denote the interior of a set A and cl(A) the closure. Find a subset of $\mathbb{R}^2$ such that as many as possible of the following sets are distinct:

$$A, \quad \text{int}(A), \quad \text{cl}(A), \quad \text{cl}(\text{int}(A)), \quad \text{int}(\text{cl}(A)), \quad \text{int}(\text{cl}(\text{int}(A))), \quad \text{cl}(\text{int}(\text{cl}(A))).$$

10. The *boundary* of a set A is the set ∂A that is the intersection of the closure of A and the closure of the complement A^c.
    (a) Show that a point p is in ∂A if and only if each neighborhood $N_\varepsilon(p)$ contains points both of A and of A^c.
    (b) Show that $\partial A = \emptyset$ if and only if A is both open and closed.
11. Is it true (always? sometimes?) that the boundary of the neighborhood $N_r(p)$ is the "sphere" $\{p : d(p, q) = r\}$?
12. Show that there is a proper subset A of $\mathbb{Q}$ with $\partial A = \emptyset$. (A subset A of a set B is *proper* if $A \neq \emptyset$ and $A \neq B$.)
13. A metric space with the property that no proper subset A has $\partial A = \emptyset$ is said to be *connected*. Show that the unit interval $[0, 1]$ is connected; show that $\mathbb{R}^n$ is connected.

6C. Coverings and Compactness

Throughout this and the remaining sections S is a metric space and other sets are subsets of S unless otherwise specified. To avoid a few uninteresting technical glitches, we assume that S is *not empty*.

Definitions. A collection $\mathcal{U}$ of subsets of S is said to *cover* a subset B of S if each point of B belongs to at least one set from the collection $\mathcal{U}$. If so, $\mathcal{U}$ is said to be *a cover* of B. If $\mathcal{U}$ is a cover and each set in the collection is open, then $\mathcal{U}$ is said to be an *open cover*.

A collection of sets $\mathcal{V}$ that consists of some or all of the sets in a cover $\mathcal{U}$ is said to be a *subcover* if it is also a cover of B.

A cover $\mathcal{U}$ of B may consist of finitely many or infinitely many sets. A *finite subcover* is a subcover that consists of only finitely many sets from $\mathcal{U}$.

Examples. 1. The collection $\mathcal{S}$ consisting of S alone is a cover for any subset B.
2. For any nonempty B, the collection $\mathcal{N}$ of all neighborhoods $N_r(p)$, $p \in B$, $r > 0$, is an open cover of B. Note that B is bounded (see below) if and only if there is a subcover that consists of a single set from $\mathcal{N}$. The collection $\mathcal{N}_1$ consisting of all $N_1(p)$, $p \in B$, is a subcover.

Definition. A set B is *compact* if each open cover of B contains a finite subcover.

Remark. It follows that a set is *not* compact if *some* cover has no finite subcover. Viewed as an adversarial procedure, to show compactness one must deal in principle with whatever clever cover one's adversary produces; to disprove compactness one gets to choose an open cover to confound the adversary. For example, choosing the open cover of $\mathbb{R}$ that consists of all intervals $(n-1, n+1)$, $n \in \mathbb{Z}$, allows one to demonstrate that $\mathbb{R}$ is not compact.

Compactness is an important concept, though its meaning and the reason for its importance become apparent only gradually. We begin by relating it to earlier concepts.

Definition. A set B is said to be *bounded* if there is a point p in S and a radius $r > 0$ such that B is a subset of $N_r(p)$.

Proposition 6.6. *If B is compact, then it is closed and bounded.*

Proof: Suppose that p belongs to the complement B^c. Let $\mathcal{U}$ be the collection $\{U_n\}$, where

$$U_n = \{q \in S : d(p,q) > 1/n\}, \quad n \in \mathbb{N}.$$

These sets are open (see Exercise 3 of Section 6B), and since p is not in B, they cover B. There is a finite subcover. But $U_1 \subset U_2 \ldots$, so the subcover contains a largest set U_n and $B \subset U_n$. Then $N_{1/n}(p)$ is included in B^c. We have proved that B^c is open, so B is closed.

Now choose any point p in S, and let $\mathcal{V}$ be the collection $\{V_n\}$, where $V_n = N_n(p)$. Then $\mathcal{V}$ is an open cover of B and there is a finite subcover. Now $V_1 \subset V_2 \subset \ldots$, so there is a largest V_n in the subcover and $B \subset V_n$. Thus B is bounded. □

Remark. The converse is not true. *In some metric spaces there are closed, bounded sets that are not compact.* We pass to a useful fact and then prove an important partial converse.

Proposition 6.7. *Any closed subset of a compact set is compact.*

Proof: Suppose that B is compact and C is a closed subset of B. The complement C^c is open. Suppose that $\mathcal{U}$ is an open cover of C. Adding C^c to this collection gives us an open cover $\mathcal{V}$ of B. There is a finite subcover, which we may assume consists of sets $U_1, \ldots, U_n$ from $\mathcal{U}$, together with C^c. Each point of C belongs to one of these sets, but not to C^c, so $U_1, \ldots, U_n$ cover C. □

The following is the partial converse to Proposition 6.6.

Theorem 6.8: Heine-Borel Theorem. *A subset of $\mathbb{R}$ or $\mathbb{C}$ that is closed and bounded is compact.*

Proof: It is enough to consider subsets of $\mathbb{C}$. If C is bounded, it is contained in a sufficiently large square with side length R centered at the origin:

$$B = \{z = x + iy : |x| \leq R/2,\ |y| \leq R/2\}.$$

If C is closed, then it is compact provided that B is compact, which we now prove. Suppose that $\mathcal{U}$ is a collection of open sets and suppose that no finite subcollection covers B. We shall see that $\mathcal{U}$ itself does not cover B. Let us say that a subset $D \subset B$ is *elusive* if no finite subcollection of $\mathcal{U}$ covers D. Thus B is assumed elusive and the goal is to chase down an elusive point of B. To do so we select a sequence of closed squares B_n. Let $B_0 = B$, with side of length R. There are four closed squares with side of length $R/2$ that cover B. Since B is elusive, at least one of these smaller squares is elusive. Choose an elusive one and denote it by B_1. Continuing, we obtain elusive closed squares

$$B_0 \supset B_1 \supset B_2 \supset \cdots; \qquad B_n \text{ has side of length } R/2^n.$$

(See Figure 4.) The centers of these squares are a Cauchy sequence $\{z_n\}$ that converges to a point z that must belong to each of the closed sets B_n. This is our elusive point. *But z cannot be so elusive*: It must belong to one of the open sets U from the collection $\mathcal{U}$. Then U would include a neighborhood of z, and that neighborhood would include B_n when n is large enough, contradicting the elusiveness of B_n. □

Note that the same idea applies to $\mathbb{R}$ and to higher dimensional spaces $\mathbb{R}^n$: In $\mathbb{R}^3$, use eight smaller closed cubes to cover a given closed cube, and so on.

Exercises

1. (a) Suppose that A is a finite subset of a metric space S. Prove that A is compact.
   (b) Suppose that the space S has the discrete metric and A is a compact subset of S. Prove that A is finite.
   (c) Use (b) to give an example of a closed, bounded set that is not compact.
2. Prove that the union of two compact subsets of a metric space is compact.

6D. Sequences, Completeness, Sequential Compactness

Many concepts from the study of real and complex sequences carry over to metric spaces. A sequence in S is, of course, a collection $\{p_n\}$ of points of S indexed by the positive or nonnegative integers. The sequence $\{p_n\}$ is said to *converge*, or to

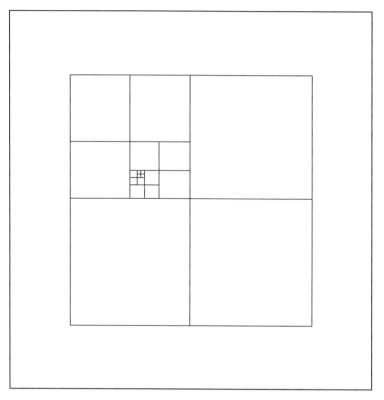

Figure 4. In search of an elusive square.

be *convergent*, if there is a point $p \in S$ such that for each $\varepsilon > 0$ there is an index N such that $n \geq N$ implies $p_n \in N_\varepsilon(p)$. The point p is unique and is said to be the *limit* of the sequence. We write $p = \lim_{n \to \infty} p_n$.

A sequence $\{p_n\}$ is said to be *bounded* if it is bounded as a set of points, that is, there is $q \in S$ and $r > 0$ such that each p_n is in $N_r(q)$. The sequence is said to be a *Cauchy sequence* if for each $\varepsilon > 0$ there is an index N such that $m, n \geq N$ implies $d(p_m, p_n) < \varepsilon$.

The following two propositions can be proved in general by the same arguments that were used in Chapter 3 to prove them in $\mathbb{C}$.

Proposition 6.9. *Each convergent sequence is a Cauchy sequence.*

Proposition 6.10. *Each Cauchy sequence is bounded.*

Definition. A metric space is said to be *complete* if each Cauchy sequence in S has a limit (in S). A subset B is said to be complete if each Cauchy sequence in B has a limit in B.

6D. Sequences, Completeness, Sequential Compactness

Theorem 3.8 can thus be restated: $\mathbb{C}$ is complete. This implies and is implied by the fact that $\mathbb{R}$ is complete.

Definition. The notion of a subsequence is defined as before. A set B is said to be *sequentially compact* if each sequence in B has a subsequence that converges to a point of B.

Theorem 6.11. *A subset of a metric space is compact if and only if it is sequentially compact.*

Partial proof: We prove the more often used part: Compactness implies sequential compactness. The other implication is left as Exercises 6–9.

Suppose that $A \subset S$ is not sequentially compact. Then there is a sequence $\{p_n\}$ in A such that no subsequence has limit in A. A consequence (Exercise 3) is that each point $p \in A$ has a neighborhood $U_p = N_{\varepsilon(p)}(p)$ with the property that there are only finitely many values of the index n such that p_n belongs to U_p. The collection $\mathcal{U}$ of these sets U_p is an open cover of A. Any finite subcollection contains p_n for only finitely many values of n. Therefore no finite subcollection covers A, so A is not compact. □

Theorems 6.8 and 6.11 combine to give another proof of the Bolzano-Weierstrass Theorem (Corollary 3.13), namely, any closed and bounded set of $\mathbb{R}^n$ or $\mathbb{C}^n$ is sequentially compact.

Exercises

1. Prove that p is in the closure of B if and only if there is a sequence in B that converges to p. (You will need to consider two cases.)
2. Suppose that the sequence $\{p_n\}$ in S has a limit p. Let $A = \{p, p_1, p_2, \ldots\}$. Prove that A is compact.
3. Show that $\mathbb{Q}$ is not complete.
4. By Exercise 2 of Section 6A, $d(x, y) = |x - y|/(1 + |x - y|)$ defines a metric on $\mathbb{R}$.
   (a) Show that $\mathbb{R}$ is bounded with respect to d.
   (b) Show that $\mathbb{R}$ is complete with respect to d.
   (c) Show that $\mathbb{R}$ is not compact with respect to d.
5. Suppose that S is complete. Show that $B \subset S$ is complete if and only if B is closed.
6. Suppose that the $\{p_n\}$ is a sequence in S and suppose that p is a point with the property that for each $\varepsilon > 0$ and each index N there is some $n > N$ such that $p_n \in N_\varepsilon(p)$. Prove that some subsequence of $\{p_n\}$ converges to p.
7. A subset B of S is said to be *totally bounded* if for each $\varepsilon > 0$ there are finitely many points $p_1, p_2, \ldots p_n$ such that the neighborhoods $N_\varepsilon(p)$ cover B. Adapt the proof of Theorem 6.8 to show that if B is complete and totally bounded, then it is compact.

8. Suppose that B is not totally bounded. Prove that there is a sequence $\{p_n\}$ in B that has no convergent subsequence.
9. Suppose that B is not complete. Prove that there is a sequence $\{p_n\}$ in B such that no subsequence converges to a point of B. Deduce from this and Exercises 6 and 7 that a set that is not compact is not sequentially compact.
10. Suppose that B is compact. Prove that it is complete and totally bounded.
11. Suppose that the sequence $\{p_n\}$ in S has the property that $\sum_1^\infty d(p_n, p_{n+1})$ is finite. Prove that $\{p_n\}$ is a Cauchy sequence.
12. (*Banach Fixed Point Theorem*, or *Contraction Mapping Theorem*) Suppose that S is a complete metric space and is not empty. Suppose that f is a function from S to itself that is a *strict contraction*, meaning that there is a constant $0 < r < 1$ such that

$$d(f(p), f(q)) \leq r \cdot d(p, q), \quad \text{all } p, q \in S.$$

Prove that there is a *fixed point*, a point p_0 such that $f(p_0) = p_0$. Prove that there is only one such point.
13. Suppose that $0 < a < 1$ and define $f(x) = (x^2 + a)/2$. Show that f has a fixed point in the interval $[0, a]$, and find it.

6E*. The Cantor Set

Let $C_0 \subset \mathbb{R}$ be the closed unit interval $[0, 1]$ and consider the sequence of sets $C_0 \supset C_1 \supset \ldots$ constructed as follows. Remove the open middle third interval from C_0, leaving the set

$$C_1 = \left[0, \tfrac{1}{3}\right] \cup \left[\tfrac{2}{3}, 1\right].$$

Remove the open middle third of each of these intervals, leaving the the set

$$C_2 = \left[0, \tfrac{1}{9}\right] \cup \left[\tfrac{2}{9}, \tfrac{1}{3}\right] \cup \left[\tfrac{2}{3}, \tfrac{7}{9}\right] \cup \left[\tfrac{8}{9}, 1\right].$$

Continue. Then C_n is the union of 2^n closed intervals, each of length $1/3^n$. The (standard) *Cantor set* is the intersection

$$C = \bigcap_{n=0}^\infty C_n = C_0 \cap C_1 \cap C_2 \cap \cdots. \tag{2}$$

Thus C is a subset of C_n for every n; put very loosely, it is what is left after the procedure of removing middle thirds of intervals is carried to completion. The total length of the intervals that make up C_n is $(2/3)^n$, which has limit 0, and the total length of the middle third interval that have been removed is

$$\frac{1}{3} + 2 \cdot \frac{1}{9} + 4 \cdot \frac{1}{27} + \cdots = \frac{1}{3}\left(\frac{2}{3} + \frac{4}{9} + \cdots\right) = 1.$$

Thus one might think that C has very few points.

6E*. The Cantor Set

Proposition 6.12. *The Cantor set C is compact and uncountable.*

Proof: Each C_n is a union of finitely many closed sets, so it is closed and the intersection C is a closed, bounded subset of $\mathbb{R}$, and hence it is compact.

Suppose that x belongs to $(0, 1]$ and consider its ternary expansion

$$x = \frac{c_1}{3} + \frac{c_2}{3^2} + \frac{c_3}{3^3} + \cdots$$

(see Section 2B). Note that $c_1 = 1$ if and only if x belongs to the interval $(1/3, 2/3]$. Suppose that $c_1 = 2$. Then $c_2 = 1$ if and only if x belongs to the interval $(7/9, 8/9]$. In fact, one can check that if there are no 1's in its ternary expansion, then x belongs to C. Conversely, if x belongs to C, then either it has no 1's in its ternary expansion or it is one of certain endpoints $2/3, 2/9, 8/9, \ldots$. Ignoring these (countably many!) exceptional points, we assign to the point with ternary expansion $.02200020220\ldots$ the point with *binary* expansion $.01100010110\ldots$, and so on. This provides a 1–1 correspondence between a subset of C and the entire uncountable interval $(0, 1]$. So C is uncountable. $\square$

Exercise

1. Discuss the result of removing the second of ten equal subintervals at each step, instead.

7
Continuous Functions

The notion of continuity of a function at a point makes sense whenever both the domain and the range of the function are metric spaces. Continuity and related issues are discussed from a general point of view in this chapter. In some circumstances, continuous functions themselves constitute a natural metric space. The Weierstrass Polynomial Approximation Theorem can be viewed in this context.

7A. Definitions and General Properties

Suppose that S and T are two sets. By a *function* f from S to T we mean an assignment to each point $p \in S$ of a unique point of T, denoted $f(p)$. The associated notation is $f : S \to T$, which is read "f is a function from S to T," or "f maps S to T."

Suppose that S and T are metric spaces and f is a function from S to T. The function f is said to be *continuous at* p if for each $\varepsilon > 0$ there is $\delta > 0$ such that $d_S(p, p') < \delta$ implies $d_T(f(p), f(p')) < \varepsilon$. The function f is said to be *continuous* if it is continuous at each point of S. If A is a subset of S and g is a function from A to T, then we extend these definitions to g by considering A itself as a metric space, with the metric it inherits from S.

From now on we take S and T to be metric spaces and derive a number of results that are exercises in the use of the various definitions.

Proposition 7.1: Continuity and sequences. *A function f from S to T is continuous at a point p in S if and only if, for each sequence p_n in S that has limit p, the sequence $\{f(p_n)\}$ has limit $f(p)$.*

Proof: Suppose first that f is continuous at p. It is a very easy exercise in the use of the definitions to show that if $\lim_{n \to \infty} p_n = p$, then $\lim_{n \to \infty} f(p_n) = f(p)$: See Exercise 1.

Conversely, suppose that f is *not* continuous at p. Then there is some $\varepsilon > 0$ with the property that for each $\delta > 0$ there is a point $p' \in N_\delta(p)$ such that $d_T(f(p), f(p')) \geq \varepsilon$. Therefore, for each $n \in \mathbb{N}$ we may choose $p_n \in N_{1/n}(p)$ such that $d_T(f(p), f(p_n)) \geq \varepsilon$. The sequence $\{p_n\}$ converges to p, but $\{f(p_n)\}$ does not converge to $f(p)$. □

Suppose that f is a function from S to T. Given any *set* $B \subset T$, we define a *set* $f^{-1}(B)$ to be the set of points of S that are taken into B by f:

$$f^{-1}(B) = \{p \in S : f(p) \in B\}. \tag{1}$$

This set is called the *inverse image* of B by f, or just the inverse image, if f is understood. Notice that f^{-1}, as defined here, is a function defined on *subsets* of T, not a function on (points of) T.

Proposition 7.2: Continuity and open sets. *A function f from S to T is continuous if and only if it is has the property: The inverse image of each open set in T is open in S.*

Proof: Suppose that f is continuous and that $B \subset T$ is open. Suppose that p is in $f^{-1}(B)$. Then $f(p)$ belongs to the open set B so there is $\varepsilon > 0$ such that $N_\varepsilon(f(p)) \subset B$. By continuity, there is $\delta > 0$ such that $p' \in N_\delta(p)$ implies $f(p') \in N_\varepsilon(f(p)) \subset B$. Therefore $N_\delta(p)$ is included in $f^{-1}(B)$. This proves that $f^{-1}(B)$ is open.

Conversely, suppose that the inverse image of each open set is open. Given $p \in S$ and $\varepsilon > 0$, let $B = N_\varepsilon(f(p))$. This is an open subset of T, so $f^{-1}(B)$ is open in S. Now p belongs to $f^{-1}(B)$, so there is $\delta > 0$ such that $N_\delta(p) \subset f^{-1}(B)$. But this means that $d_T(f(p), f(p')) < \varepsilon$ when $d_S(p, p') < \delta$, so f is continuous at p. □

If f is a function from S to T and g is a function from T to U, the *composition* of f and g is the function denoted by $g(f)$ or $g \circ f$:

$$g(f) : S \to U, \qquad [g(f)](p) = g(f(p)), \qquad p \in S.$$

Proposition 7.3: Composition of continuous functions. *Suppose that S, T, and U are metric spaces and that the functions f mapping S to T and g mapping T to U are continuous. Then the composition $g(f)$ is continuous from S to U.*

Proof: Suppose that C is an open subset of U. Then $B = g^{-1}(C)$ is open in T, and so $f^{-1}(B) = f^{-1}(g^{-1}(B))$ is open in S. It is easy to check that

$$\bigl(g(f)\bigr)^{-1}(C) = f^{-1}\bigl(g^{-1}(C)\bigr).$$

Thus inverse images of open sets under $g(f)$ are open and so $g(f)$ is continuous. □

Definition. A function f from S to T is said to be *bounded* if the image set $f(S) = \{f(p) : p \in S\}$ is a bounded subset of T.

Theorem 7.4: Continuity and compactness. *Suppose that A is a compact subset of S and that f is a continuous function from A to T. Then the image $f(A) = \{f(p) : p \in A\}$ is a compact subset of T. In particular, f is bounded.*

Proof: Suppose that $\mathcal{V}$ is an open cover of $f(A)$. Let $\mathcal{U}$ be the collection of subsets of S consisting of the inverse images of the sets in the collection $\mathcal{V}$. Proposition 7.2 implies that these sets are open. Given any point p in A, $f(p)$ belongs to some V from the collection $\mathcal{V}$, so p belongs to the set $f^{-1}(V)$ from the collection $\mathcal{U}$. Thus $\mathcal{U}$ is an open cover of A. Let $\{U_1, \ldots, U_n\}$ be a finite subcover. Then each U_j is $f^{-1}(V_j)$ for some V_j in the collection $\mathcal{V}$, and the collection $\{V_j\}$ is a finite subcover of $f(A)$. □

Definition. A function f from S to T is said to be *uniformly continuous* if for each $\varepsilon > 0$ there is a $\delta > 0$ such that $d_S(p, p') < \delta$ implies that $d_T\bigl(f(p), f(p')\bigr) < \varepsilon$.

Continuity says that, for each p and each $\varepsilon > 0$, there is $\delta > 0 \ldots$. Uniform continuity says that, for each $\varepsilon > 0$, there is $\delta > 0 \ldots$. It is obvious from the definitions that uniform continuity implies continuity. The converse is not true.

Examples. 1. The function from $\mathbb{R}$ to $\mathbb{R}$ defined by $f(x) = x^2$ is continuous but is not uniformly continuous. In fact, given any $\delta > 0$, suppose that $|y - x| = \delta/2 < \delta$. Then

$$|f(y) - f(x)| = |y^2 - x^2| = |(y-x)(y+x)| = \tfrac{1}{2}\delta \, |y + x|,$$

which is larger than 1 whenever $y > x > 1/\delta$.

2. The function from $(0, 1)$ to $\mathbb{R}$ defined by $g(x) = 1/x$ is continuous but not uniformly continuous. Points x, y that are very close to 0 (and thus close to each other) can be sent to points $g(x), g(y)$ that are very far apart.

These examples show that the following theorem contains useful information.

Theorem 7.5: Compactness and uniform continuity. *If A is compact and f is a continuous function from A to T, then f is uniformly continuous.*

Proof: Given $\varepsilon > 0$, we create an open cover of A as follows. For each p in A, choose $\delta(p) > 0$ such that if p' is in the neighborhood $V_p = N_{2\delta(p)}(p)$, then $d_T(f(p), f(p')) < \varepsilon/2$. Let $\mathcal{U}$ be the collection of neighborhoods $U_p = N_{\delta(p)}(p)$. There is a finite subcover, corresponding to points $p_1, \ldots, p_n$. Let δ be the smallest of the numbers $\delta(p_j)$. Suppose now that $d_S(p, p') < \delta$. Choose p_j so that p belongs to U_{p_j}. The triangle inequality and the choices we made imply that both p and p' are in V_{p_j}. Therefore

$$d_T(f(p), f(p')) \leq d_T(f(p), f(p_j)) + d_T(f(p_j), f(p')) < \frac{\varepsilon}{2} + \frac{\varepsilon}{2} = \varepsilon. \quad \square$$

Exercises

1. Prove that if f is continuous at p and $\lim_{n \to \infty} p_n = p$, then $\lim_{n \to \infty} f(p_n) = f(p)$.
2. Prove that $f : S \to T$ is continuous if and only if the inverse image of each closed set is closed.
3. Suppose that $f : S \to T$ and $g : T \to U$. Suppose that f is continuous at the point $p \in S$ (but is not assumed to be continuous elsewhere) and suppose that g is continuous at the point $q = f(p)$. Prove that $g(f)$ is continuous at p.
4. Suppose that a is a point of the metric space S. Define $g(p) = d(a, p)$, $p \in S$. Prove that g is *uniformly* continuous.
5. Suppose that A is an unbounded subset of S. Show that there is a uniformly continuous, real-valued function on A that is not bounded.
6. Suppose that B is a subset of S that is not closed. Show that there is a bounded, uniformly continuous real-valued function on B that does not attain a minimum.
7. Suppose that B is a subset of S that is not closed. Show that there is a continuous, real-valued function on B that is not bounded.
8. Suppose that B is a subset of S that is not closed. Show that there is a continuous real-valued function defined on B that is not uniformly continuous.
9. Find a bounded, continuous function $f : \mathbb{R} \to \mathbb{R}$ such that $f(\mathbb{R})$ is neither open nor closed.
10. Find a closed, bounded subset B of $\mathbb{Q}$ and a continuous, real-valued function f defined on B such that f is not bounded.
11. Suppose that S has the discrete metric. Show that every function from S to a metric space is uniformly continuous.
12. Find an unbounded set A such that every function from A to a metric space is uniformly continuous.
13. Suppose that A is a nonempty subset of the metric space S. Define the distance from a point p of S to the set A to be

$$d_A(p) = \inf\{d(p, q) : q \in A\}.$$

Prove that $d_A(p) = 0$ if and only if p is in the closure of A. Prove that d_A is a uniformly continuous function on S.

14. Suppose that f is a real-valued uniformly continuous function on a set A.
    (a) Suppose that $\{p_n\}_1^\infty$ is a Cauchy sequence in A. Show that $\{f(p_n)\}_1^\infty$ is a Cauchy sequence in $\mathbb{R}$.
    (b) Show that there is a continuous real-valued function g defined on the closure of A such that, for each p in A, $g(p) = f(p)$. Show that g is unique.
15. Suppose that A is a dense subset of the metric space S, and suppose that $f : A \to \mathbb{R}$ is uniformly continuous. Prove that there is a unique continuous function $g : S \to \mathbb{R}$ such that $g = f$ on A.

7B. Real- and Complex-Valued Functions

If f and g are functions from S to $\mathbb{C}$, we define the sum and difference functions $f \pm g$ and the product function fg by

$$[f \pm g](p) = f(p) \pm g(p), \qquad [fg](p) = f(p)\,g(p), \qquad p \in S.$$

If $c \in \mathbb{C}$, we define cf by $[cf](p) = c \cdot f(p)$, $p \in S$. The function $|f|$ is defined to have value $|f(p)|$ at $p \in S$. If $g(p) \neq 0$, all $p \in S$, we define the quotient f/g by $[f/g](p) = f(p)/g(p)$, $p \in S$.

Functions that take real values are a special case of the preceding.

Proposition 7.6: Continuity and algebra. *Suppose that f and g are continuous functions from S to $\mathbb{C}$ and suppose that c is complex. Then the functions cf, $|f|$, $f \pm g$, and fg are continuous. If $g(p) \neq 0$, all $p \in S$, then f/g is continuous.*

Proof: It is enough to prove that each of these functions is continuous at each point $p \in S$. According to Proposition 7.1, it is enough to examine convergent sequences $\{p_n\}$, $\lim_{n \to \infty} p_n = p$. Given such a sequence, we may define $a_n = f(p_n)$, $b_n = g(p_n)$ and apply Theorem 3.9 on algebraic properties of limits. □

Remark: The function $f : \mathbb{C} \to \mathbb{C}$ defined by $f(z) = z$, all z, is continuous: Given $\varepsilon > 0$, take $\delta = \varepsilon$. It follows from Proposition 7.6 that any polynomial function $h(z) = \sum_0^n a_n z^n$ is continuous. [On the other hand, we proved in Chapter 5 that any function defined by a convergent power series is differentiable; a similar but simpler proof shows that it is continuous. A polynomial is a power series with only finitely many nonzero coefficients, so the continuity of polynomials is a very special case.]

Theorem 7.7: Maximum and minimum. *Suppose that A is a compact subset of S and that f is a continuous function from A to $\mathbb{R}$. Then f attains its minimum and maximum values: There are points $p_{\min}$, $p_{\max}$ in A such that*

$$f(p_{\min}) \leq f(p) \leq f(p_{\max}), \qquad \text{all } p \in A. \tag{2}$$

Proof: According to Theorem 7.4, the image $f(A)$ is a compact subset of $\mathbb{R}$; therefore, it is closed and bounded. A closed, bounded set in $\mathbb{R}$ contains its least upper bound (Exercise 7 of Section 6B). Thus there is a point $p_{\max}$ in A such that $f(p_{\max})$ is the least upper bound of $f(A)$. Similarly, there is a point $p_{\min}$ in A such that $f(p_{\min})$ is the greatest lower bound. □

Exercises

1. Suppose that f is a real-valued function on $\mathbb{R}$ that is *additive*: For any real x, y, $f(x+y) = f(x) + f(y)$. Prove that if f is continuous, then there is a constant a such that $f(x) = ax$ for all real x.
2. Suppose that f is a continuous real-valued function on $\mathbb{R}$, and suppose that for any x and y in $\mathbb{R}$,
$$f(x+y) + f(x-y) = 2\bigl[f(x) + f(y)\bigr].$$
Prove that there is a constant a such that $f(x) = a x^2$ for all real x.
3. A real-valued function f defined on an open interval (a, b) is said to be *convex* if
$$f\bigl(tx + (1-t)y\bigr) \le tf(x) + (1-t)f(y)$$
for all $x, y \in (a, b)$ and each $t \in (0, 1)$. Prove that any convex function is continuous.
4. Prove that any increasing convex function of a convex function is convex.
5. Suppose that f is a continuous real-valued function defined on an open interval (a, b), and suppose that
$$f\left(\frac{x+y}{2}\right) \le \frac{f(x) + f(y)}{2}, \quad \text{all } x, y \in (a, b).$$
Prove that f is convex.
6. Suppose that the power series $\sum_0^\infty a_n z^n$ has radius of convergence $R > 0$. Let A be the disk $\{z \in \mathbb{C} : |z| < R\}$. Give two proofs that $f(z) = \sum_0^\infty a_n z^n$ is a continuous function on A.

7C. The Space $C(I)$

Suppose that A is a compact set in a metric space S. Let $C(A; \mathbb{C})$ denote the set whose elements (points) are the continuous functions from A to $\mathbb{C}$. Similarly, $C(A; \mathbb{R})$ denotes the set whose elements are the continuous functions from A to $\mathbb{R}$. For convenience we use the notation $C(A)$ to denote either space of functions.

We look first at the following special case:

$$I = [a, b] = \{x \in \mathbb{R} : a \le x \le b\};$$
$$C(I) = C(I, \mathbb{R}) = \{f : f : I \to \mathbb{R}; \ f \text{ continuous}\}. \tag{3}$$

(We assume that $a < b$, so the interval I is not trivial.) A "point" of $C(I)$ is a continuous real-valued function f on I. Such a function may be visualized by identifying it with its graph, pictured as a continuous curve in the plane.

The standard metric in $C(I)$ is obtained by starting with the norm

$$\|f\| = \sup\{|f(x)| : x \in I\}. \tag{4}$$

This is well defined and finite, by Theorem 7.4. Then define

$$d(f, g) = \|f - g\| = \sup\{|f(x) - g(x)| : x \in I\}. \tag{5}$$

It is not difficult to check that the norm $\|f\|$ has the properties, listed in Section 6A, that define a norm:

$$\|f\| \geq 0, \quad \|f\| = 0 \iff f \equiv 0; \tag{6}$$

$$\|cf\| = |c| \cdot \|f\|, \quad c \in \mathbb{R}; \tag{7}$$

$$\|f + g\| \leq \|f\| + \|g\|. \tag{8}$$

The distance between f and g is precisely the maximum vertical distance between their graphs.

A metric can be defined in the same way in $C(A; \mathbb{C})$ or $C(A; \mathbb{R})$:

$$\|f\| = \sup\{|f(p)| : p \in A\}; \quad d(f, g) = \|f - g\|.$$

The single most important fact about these spaces of functions is the following.

Theorem 7.8. *For any compact A, the spaces $C(A; \mathbb{R})$ and $C(A; \mathbb{C})$ are complete.*

Proof: Suppose that $\{f_n\}$ is a Cauchy sequence in $C(A; \mathbb{C})$. (The proof is exactly the same for $C(A; \mathbb{R})$.) For each $p \in A$, the sequence $\{f_n(p)\}$ is a Cauchy sequence of complex numbers, in fact,

$$|f_n(p) - f_m(p)| \leq \|f_n - f_m\| = d(f_n, f_m).$$

Given any $\varepsilon > 0$, we can choose N so large that $n, m \geq N$ implies $d(f_n, f_m) < \varepsilon$. Therefore we may define $f(p) = \lim_{n \to \infty} f_n(p)$. For each p and each m,

$$|f(p) - f_m(p)| = \lim_{n \to \infty} |f_n(p) - f_m(p)| \leq \varepsilon$$

if $m \geq N$, so $d(f, f_m) \leq \varepsilon$ if $m \geq N$. This would complete the proof – except that it is necessary to show that f is continuous. To do so, we argue as follows. Given any $\varepsilon > 0$, choose N as before. The function f_N is uniformly continuous, so there is $\delta > 0$ such that $d_S(p, q) < \delta$ implies $|f_N(p) - f_N(q)| < \varepsilon$. Suppose that

$d_S(p, q) < \delta$. Then, by what has already been shown,

$$|f(p) - f(q)| \leq |f(p) - f_N(p)| + |f_N(p) - f_N(q)| + |f_N(q) - f(q)|$$
$$\leq \varepsilon + \varepsilon + \varepsilon = 3\varepsilon.$$

Therefore f is (uniformly) continuous. □

Example. Let $I = [0, 1]$ and let f_n in $C(I), n = 1, 2, \ldots$ be defined by $f_n(x) = x^n$, $x \in I$. Then $f_n(x)$ has limit $f(x) = 0$ for $0 \leq x < 1$ and $f(x) = 1$ for $x = 1$. This limit function is not continuous – but this does not contradict Theorem 7.8. (Why?)

We know that any real number can be approximated by rationals. More precisely, given a real x and any $\varepsilon > 0$, we can find a rational r (for example, a partial sum of the decimal, or binary, or ternary expansion of x) such that $|x - r| < \varepsilon$. This gives one the comforting feeling that the reals are not so mysterious after all. Elements of $C(I)$, viewed simply as all possible graphs, with all sorts of cusps and corners, may seem yet more mysterious, but they too can be approximated arbitrarily closely by less exotic objects.

Theorem 7.9: Weierstrass Polynomial Approximation Theorem. *Suppose that a and b are real, $a < b$, and suppose that f is a continuous real-valued function on $[a, b]$. Given any $\varepsilon > 0$, there is a polynomial P such that $\|f - P\| < \varepsilon$:*

$$|f(x) - P(x)| < \varepsilon, \quad \text{all } x, a \leq x \leq b.$$

There are a number of ways to prove this theorem. One is given in the next section.

A subset A of S is said to be *dense* in S if S is the closure of A. This is the same as saying that, for any point p in S and any $\varepsilon > 0$, there is a point $q \in A$ that belongs to $N_\varepsilon(p)$. Thus $\mathbb{Q}$ is dense in $\mathbb{R}$. Theorem 7.9 says: *Polynomials are dense in the space of real-valued continuous functions on $[a, b]$.*

Remark. The notion of convergence of a sequence of real- or complex-valued functions that corresponds to the metric can be generalized beyond compact sets and continuous functions. If S is any set and if $\{f_n\}_0^\infty$ and f are functions from S to R or to $\mathbb{C}$, then the sequence $\{f_n\}$ is said to *converge uniformly* to the function f if for each $\varepsilon > 0$ there is an index N such that $n \geq N$ implies

$$|f_n(p) - f(p)| < \varepsilon, \quad \text{all } p \in S.$$

Thus a sequence $\{f_n\}$ in the space $C(A)$ converges to f with respect to the metric (4) if and only if it converges uniformly to f.

In particular, it is easy to see that the Weierstrass Polynomial Approximation Theorem may be reformulated in the following way: *If f is a continuous real-valued function on $[a, b]$, there is a sequence $\{P_n\}$ of polynomials that converges to f uniformly on $[a, b]$.*

Exercises

1. Which of the following sequences of functions f_1, f_2, f_3, ... converges uniformly on the interval $[0, 1]$?

    (a) $\qquad f_n(x) = nx^2(1-x)^n.$
    (b) $\qquad f_n(x) = n^2 x(1-x^2)^n.$
    (c) $\qquad f_n(x) = n^2 x^3 e^{-nx^2}.$
    (d) $\qquad f_n(x) = \dfrac{x^2}{x^2 + (1-nx)^2}.$

2. Let $I = [0, 1]$ and $A = \{f \in C(I) : |f(x)| \leq 1,\ \text{all}\ x \in I\}$.
    (a) Show that A is closed and bounded.
    (b) Show that the sequence $(f_n)_{n=0}^{\infty}$ with $f_n(x) = x^n$ is a sequence in A that has no convergent subsequence. (In fact, for fixed m the distance $d(f_n, f_m)$ increases as $n \to \infty$.)

3. Let I and A be the same as in the preceding exercise. Let $U_n = \{f \in C(I) : |f(0) - f(1/n)| < 1\}$, $n \in \mathbb{N}$. Show that $\{U_n\}$ is an open cover of A but there is no finite subcover.

4. Let $\mathcal{P}_n$ denote the subspace of $C(I)$ consisting of functions that are polynomials of degree $\leq n$.
    (a) Prove, for $n = 0$, $n = 1$, and $n = 2$, that this is a closed subset.
    (b) Give a strategy for proving the result for general n.

5. Show that functions whose graphs are polygonal lines in $\mathbb{R}^2$ are dense in $C(I)$.

6. A metric space is said to be *separable* if it has a countable dense subset. Thus $\mathbb{R}$ is separable, since $\mathbb{Q}$ is countable and dense.
    (a) Prove that $\mathbb{C}$ is separable.
    (b) Prove that $C(I)$ is separable.

7. Consider the set of all real polynomials that have only terms of *even degree*, for example, $x^6 - 3x^2 + 7$, but not $x + 2$ or $2x^4 - x^3 + 4$.
    (a) Prove that these polynomials are dense in $C(I)$ if $I = [0, 1]$.
    (b) Is this true when $I = [-1, 1]$?

8. Define a sequence of polynomials P_0, P_1, P_2, ... by $P_0(x) \equiv 0$ and
$$P_{n+1}(x) = P_n(x) + \frac{x^2 - (P_n(x))^2}{2}, \qquad n = 0, 1, 2, \ldots.$$
Prove that this sequence converges uniformly to the function $|x|$ on the interval $[-1, 1]$.

7D*. Proof of the Weierstrass Polynomial Approximation Theorem

The following proof is due to S. N. Bernstein. To simplify matters slightly, we observe that we may take the interval I to be the interval $[0, 1]$. In fact, given $a < b$, let

$$\varphi(t) = (1-t)a + tb, \quad 0 \leq t \leq 1; \quad \psi(x) = \frac{(x-a)}{(b-a)}, \quad a \leq x \leq b.$$

Then f is a continuous function on $[a, b]$ if and only if $g = f(\varphi)$ is a continuous function on $[0, 1]$. If polynomials $\{P_n\}$ converge uniformly to g on $[0, 1]$, then the functions $Q_n = P_n(\psi)$ are polynomials that converge to $f = g(\psi)$ on $[a, b]$.

Thus, suppose that f is a continuous real-valued function on $I = [0, 1]$. We define a sequence of polynomials as follows:

$$P_n(x) = \sum_{k=0}^{n} f\left(\frac{k}{n}\right) \binom{n}{k} x^k (1-x)^{n-k}, \quad n \in \mathbb{N}. \tag{9}$$

(See Figure 5.) As usual, $\binom{n}{k}$ denotes the binomial coefficient $n!/k!\,(n-k)!$. Note that P_n is a polynomial of degree $\leq n$, because each of the functions $x^k(1-x)^{n-k}$ is a polynomial of degree n.

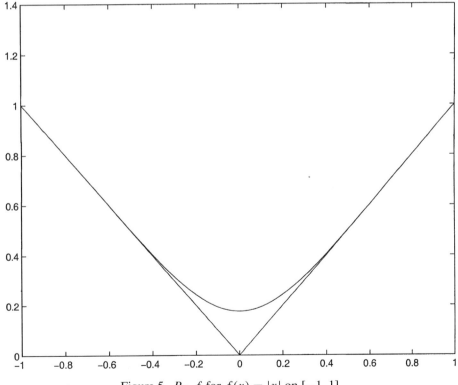

Figure 5. $P_{20} f$ for $f(x) = |x|$ on $[-1, 1]$.

These polynomials have a nice probabilistic interpretation, which is discussed below. Let us show that they converge uniformly to f.

A key step is to compute these polynomials in three cases: the cases $f(x) = x^v$, $x \in I$, where $v = 0, 1, 2$. These lead to the three sums evaluated next.

Lemma 7.10. *For each real x*

$$\sum_{k=0}^{n} \binom{n}{k} x^k (1-x)^{n-k} = 1; \tag{10}$$

$$\sum_{k=0}^{n} k \binom{n}{k} x^k (1-x)^{n-k} = nx; \tag{11}$$

$$\sum_{k=0}^{n} k^2 \binom{n}{k} x^k (1-x)^{n-k} = n(n-1)x^2 + nx. \tag{12}$$

Proof: The identity (10) is immediate, because the left side is just the binomial expansion of $[x + (1-x)]^n$. To verify (11), notice that

$$k \binom{n}{k} = \frac{n!}{(k-1)!(n-k)!} = \frac{n \cdot (n-1)!}{(k-1)!(n-1-(k-1))!} = n \binom{n-1}{k-1}. \tag{13}$$

Note also that we may take the sum in (11) starting from $k = 1$. Let $j = k - 1$ and use (13) to rewrite the left-hand side of (11) as

$$\sum_{j=0}^{n-1} n \binom{n-1}{j} x^{j+1} (1-x)^{n-1-j} = nx[x + (1-x)]^{n-1} = nx.$$

Note that $k^2 = k(k-1) + k$, so the final identity (12) can be deduced immediately from (11) together with

$$\sum_{k=0}^{n} k(k-1) \binom{n}{k} x^k (1-x)^{n-k} = n(n-1)x^2. \tag{14}$$

To obtain (14) we proceed as before. As in (13),

$$k(k-1) \binom{n}{k} = n(n-1) \binom{n-2}{k-2}.$$

The sum in (14) may be taken starting with $k = 2$ and written with $j = k - 2$ to convert it to $n(n-1)x^2[x + (1-x)]^2$. □

To prove uniform convergence we need to investigate the difference $P_n(x) - f(x)$, $x \in I$. Now $P_n(x)$ is the sum (9). We take advantage of (10) to write

7D*. Proof of the Weierstrass Polynomial Approximation Theorem

$f(x) = f(x) \cdot 1$ as a sum and obtain

$$P_n(x) - f(x) = \sum_{k=0}^{n} f\left(\frac{k}{n}\right) \binom{n}{k} x^k (1-x)^{n-k} - \sum_{k=0}^{n} f(x) \binom{n}{k} x^k (1-x)^{n-k}$$

$$= \sum_{k=0}^{n} \left[f\left(\frac{k}{n}\right) - f(x) \right] \binom{n}{k} x^k (1-x)^{n-k}. \tag{15}$$

For any given (small) $\delta > 0$ let us break the last sum into two parts:

$$\sum_{|x-k/n|<\delta} \left[f\left(\frac{k}{n}\right) - f(x) \right] \binom{n}{k} x^k (1-x)^{n-k}$$

$$+ \sum_{|x-k/n|\geq\delta} \left[f\left(\frac{k}{n}\right) - f(x) \right] \binom{n}{k} x^k (1-x)^{n-k}. \tag{16}$$

We shall see that the first sum in (16) is small if δ is small, because of continuity. For fixed $\delta > 0$, the second is small when n is large, because the weights $\binom{n}{k} x^k (1-x)^{n-k}$ are small when k/n is not close to x. To quantify this, suppose that $\varepsilon > 0$ is given. Choose $\delta > 0$ so small that, for $x, y \in I$, $|x - y| < \delta$ implies $|f(x) - f(y)| < \varepsilon/2$. Then the first sum in (16) has absolute value less than

$$\frac{\varepsilon}{2} \cdot \sum_{k=0}^{n} \binom{n}{k} x^k (1-x)^{n-k} = \frac{\varepsilon}{2}.$$

Now $|f(x)| \leq \|f\|$, all $x \in I$, so the second sum in (16) has absolute value at most

$$2\|f\| \sum_{|x-k/n|\geq\delta} \binom{n}{k} x^k (1-x)^{n-k}. \tag{17}$$

We fix δ and turn to the second sum in (16). Now we use the fact that $\delta^{-2}(x - k/n)^2$ is nonnegative for all k and is ≥ 1 for the values of k that occur in the sum (17). Therefore

$$\sum_{|x-k/n|\geq\delta} \binom{n}{k} x^k (1-x)^{n-k} \leq \frac{1}{\delta^2} \sum_{k=0}^{n} \left(x - \frac{k}{n}\right)^2 \binom{n}{k} x^k (1-x)^{n-k}$$

$$= \frac{1}{\delta^2} \sum_{k=0}^{n} \left(x^2 \cdot 1 - \frac{2x}{n} \cdot k + \frac{1}{n^2} \cdot k^2\right) \binom{n}{k} x^k (1-x)^{n-k}$$

$$= \frac{1}{\delta^2} \left[x^2 - \frac{2x}{n} \cdot (nx) + \frac{1}{n^2} \cdot (n(n-1)x^2 + nx) \right]$$

$$= \frac{x(1-x)}{n\delta^2} \leq \frac{1}{4n\delta^2} \tag{18}$$

for $x \in I$; we have made use of (10)–(12). Therefore, the absolute value of (16) is less than ε for every $x \in I$, provided that $n \geq \varepsilon/2 \|f\| \delta^2$. We have proved that the polynomials P_n converge uniformly to f on the interval I.

Going back to a function f that is defined and continuous on an interval $[a, b]$, one can see from the remarks at the beginning of this section that the corresponding Bernstein polynomials that converge uniformly to f on $[a, b]$ are

$$P_n(x) = \sum_{k}^{n} \binom{n}{k} f\left(a + \frac{k}{n}(b-a)\right) \left(\frac{x-a}{b-a}\right)^k \left(\frac{b-x}{b-a}\right)^{n-k}. \quad (18)$$

Here is the promised probabilistic interpretation. Fix $x \in I$. Suppose that one has a coin that has probability x of coming up heads and probability $1 - x$ of coming up tails when tossed. The probability of any particular string of n tosses that result in k heads is $x^k(1-x)^{n-k}$. The number of ways that k heads can come up in n tosses is $\binom{n}{k}$. Therefore, the probability of k heads in n tosses is $\binom{n}{k} x^k (1-x)^{n-k}$. Thus (10) gives the probability of getting *some* number of heads ($0 \leq k \leq n$) in n tosses, while (11) gives the expected number of heads in n tosses – that is, the expected average over many repetitions of n tosses.

The inequality (18) gives an estimate of how little likelihood there is that the *proportion of heads* will differ from the probability x by more than δ and shows that this becomes less and less likely as n increases.

Finally, suppose that the coin is to be tossed n times and there will be a payoff of $f(k/n)$ if the number of heads is k. Then the expected payoff is $P_n(x)$. This is why one could expect $P_n(x)$ to be close to $f(x)$ if n is large.

Exercise

1. (a) Compute the Bernstein polynomials P_n (see (9)) for the function $f(x) = e^x$ on $[0, 1]$.
   (b) Show that for any $\varepsilon > 0$ it is true for large enough n that

$$\left(1 + \frac{x}{n}\right)^n \leq P_n(x) \leq \left(1 + \frac{x+\varepsilon}{n}\right)^n, \quad 0 \leq x \leq 1.$$

8
Calculus

In this chapter we present a rapid review of the theoretical foundations of differential and integral calculus in one variable. This includes the Mean Value Theorem and the Generalized Mean Value Theorem, various versions of the Fundamental Theorem, and Taylor expansions.

8A. Differential Calculus

Suppose that f is a real- or complex-valued function defined in a neighborhood of a point $x \in \mathbb{R}$. Then f is said to be *differentiable* at x with *derivative* $f'(x)$ if for each $\varepsilon > 0$ there is $\delta > 0$ such that

$$\left| \frac{f(y) - f(x)}{y - x} - f'(x) \right| < \varepsilon \quad \text{if } 0 < |y - x| < \delta. \tag{1}$$

This can be rewritten somewhat informally as

$$\lim_{y \to x} \frac{f(y) - f(x)}{y - x} = f'(x). \tag{2}$$

Remarks

1. Differentiability at x implies continuity at x, as is easily seen by multiplying (1) by $|y - x|$.
2. A complex-valued function is differentiable at x if and only if its real and imaginary parts are differentiable at x. Thus, considering complex-valued functions does not introduce any genuine complications here, and it will be convenient later.

Proposition 8.1: Differentiation and algebra. *Suppose that f and g are real- or complex-valued functions defined in a neighborhood of the point x in $\mathbb{R}$, and suppose that c is complex. If f and g are both differentiable at x, then cf, $f \pm g$,* *and fg are differentiable at x and*

$$[cf]'(x) = c \cdot f'(x); \tag{3}$$

$$[f \pm g]'(x) = f'(x) \pm g'(x); \tag{4}$$

$$[fg]'(x) = f(x)g'(x) + f'(x)g(x). \tag{5}$$

If $g(x) \neq 0$, then $1/g$ is defined in a neighborhood of x and

$$\left[\frac{1}{g}\right]'(x) = -\frac{g'(x)}{g(x)^2}. \tag{6}$$

Proof: The identities (3) and (4) are very easy. The usual trick of adding and subtracting a term leads to

$$\frac{f(y)g(y) - f(x)g(x)}{y - x} = f(y) \left[\frac{g(y) - g(x)}{y - x}\right] + \left[\frac{f(y) - f(x)}{y - x}\right] g(x),$$

which leads to (5), Leibniz's rule.

By assumption, g is continuous at x, so $g(x) \neq 0$ implies that g is nowhere zero in some neighborhood of x, and thus $1/g$ is defined there. Now

$$\frac{1/g(y) - 1/g(x)}{y - x} = -\frac{1}{g(y)g(x)} \cdot \frac{g(y) - g(x)}{y - x},$$

which leads to (6). □

In what follows we shall make use of Proposition 8.2 in conjunction with the (obvious) facts: $f(x) = constant$ implies $f' = 0$, while $f(x) = x$ for all x implies $f' = 1$.

Proposition 8.2: Chain rule. *Suppose that g is real-valued and that the composition $f(g)$ is defined in a neighborhood of $x \in \mathbb{R}$. Suppose that g is differentiable at x and that f is differentiable at $g(x)$. Then $f(g)$ is differentiable at x and $[f(g)]'(x) = f'(g(x)) g'(x)$.*

Proof: Differentiability of f at $g(x)$ implies that

$$f(g(y)) - f(g(x)) = [f'(g(x)) + r(y)] \cdot [g(y) - g(x)],$$

where the error term $r(y)$ has limit 0 as $g(y)$ approaches $g(x)$. Since g is continuous at x, we conclude that $r(y)$ has limit 0 as $y \to x$. Therefore the difference quotient

$$\frac{f(g(y)) - f(g(x))}{y - x} = [f'(g(x)) + r(y)] \frac{g(y) - g(x)}{y - x}$$

has limit $f'(g(x)) g'(x)$ as $y \to x$. □

8A. Differential Calculus

Theorem 8.3: Rolle's Theorem. *Suppose that f is real-valued and continuous on the closed interval $[a, b] \subset \mathbb{R}$ and differentiable at every point of the open interval (a, b). Suppose also that $f(a) = f(b)$. Then there is a point $c \in (a, b)$ such that $f'(c) = 0$.*

Proof: If f is constant on the interval, then f' vanishes at every point of the open interval. Because $[a, b]$ is compact, we know that f attains it minimum and maximum values. If f is not constant, then either the maximum value is larger than $f(a)$ or the minimum value is smaller than $f(a)$, or both. Suppose that the maximum value is larger than $f(a)$. Then any point c at which it is attained lies in (a, b). Let c be such a point. The numerator of the difference quotient

$$\frac{f(y) - f(c)}{y - c}, \quad y \in [a, b]$$

is always ≤ 0 and the denominator can have either sign. We have assumed the existence of the limit $f'(c)$ as $y \to c$, so the only possible limit is 0. The same argument applies if f attains a minimum $< f(a)$. □

Theorem 8.4: Mean Value Theorem. *Suppose that f is real-valued and continuous on the closed interval $[a, b] \subset \mathbb{R}$ and differentiable at every point of the open interval (a, b). Then there is a point $c \in (a, b)$ such that*

$$\frac{f(b) - f(a)}{b - a} = f'(c). \tag{7}$$

Proof: Let g be the function defined on $[a, b]$ by

$$g(x) = f(x) - \frac{f(b) - f(a)}{b - a}(x - a).$$

Note that $g(a) = g(b) = f(a)$. By Rolle's Theorem there is a point $c \in (a, b)$ such that

$$0 = g'(c) = f'(c) - \frac{f(b) - f(a)}{b - a}. \quad □$$

The Mean Value Theorem (MVT) is fundamental for the theory of calculus. The reader who doubts this is invited to consider the following corollaries and find rigorous proofs that do *not* make use of the MVT. Note that the MVT implies Rolle's Theorem; on the other hand, we used Rolle's Theorem to prove the MVT.

Corollary 8.5. *If $f'(x) = 0$, all x in the interval (a, b), then f is constant on (a, b).*

Proof: It is enough to prove this for the real and imaginary parts of a function, so suppose that f is real. Given x and y in (a, b), with $x < y$, apply the MVT to f on the interval $[x, y]$ to conclude that $f(y) - f(x) = 0$. □

Definition. A real-valued function f on the interval (a, b) is said to be

nondecreasing if $x, y \in (a, b)$, $x < y$ implies $f(x) \leq f(y)$;
strictly increasing if $x, y \in (a, b)$, $x < y$ implies $f(x) < f(y)$;
nonincreasing if $x, y \in (a, b)$, $x < y$ implies $f(x) \geq f(y)$;
strictly decreasing if $x, y \in (a, b)$, $x < y$ implies $f(x) > f(y)$.

A function that is either nonincreasing or nondecreasing is said to be *monotone*; a function that is either strictly increasing or strictly decreasing is said to be *strictly monotone*.

It follows easily from the definitions that if f is nondecreasing and is differentiable at x, then $f'(x) \geq 0$; if f is nonincreasing and differentiable at x, then $f'(x) \leq 0$. The converse results follow from the MVT.

Corollary 8.6. *Suppose that the real-valued function f on the interval (a, b) is differentiable at each point of (a, b). Then*

$f'(x) \geq 0$, all $x \in (a, b)$ implies that f is nondecreasing;
$f'(x) > 0$, all $x \in (a, b)$ implies that f is strictly increasing;
$f'(x) \leq 0$, all $x \in (a, b)$ implies that f is nonincreasing;
$f'(x) < 0$, all $x \in (a, b)$ implies that f is strictly decreasing.

The following generalization of the MVT is also useful.

Theorem 8.7: Generalized Mean Value Theorem. *Suppose that f and g are continuous real-valued functions on the interval $[a, b]$ that are differentiable at every point of (a, b). Suppose also that $g'(x) \neq 0$, all $x \in (a, b)$. Then there is a point $c \in (a, b)$ such that*

$$\frac{f(b) - f(a)}{g(b) - g(a)} = \frac{f'(c)}{g'(c)}. \tag{8}$$

Proof: Notice that MVT and the assumption that $g' \neq 0$ on (a, b) imply that $g(b) - g(a) \neq 0$, so the left hand side of (8) is well-defined. Define $h : [a, b] \to \mathbb{R}$ by

$$h(x) = f(x)\big[g(b) - g(a)\big] - g(x)\big[f(b) - f(a)\big].$$

Then $h(a) = f(a)g(b) - f(b)g(a) = h(b)$, so by Rolle's Theorem there is $c \in (a, b)$ such that

$$0 = h'(c) = f'(x)[g(b) - g(a)] - g'(x)[f(a) - f(b)];$$

divide by $[g(b) - g(a)]g'(x)$. □

Note that MVT is the case $g(x) = x$ of the Generalized Mean Value Theorem (GMVT). The GMVT is the basis for a well-known method for calculating limits.

Corollary 8.8: L'Hôpital's Rule. *Suppose that the real-valued functions f and g are differentiable at each point of the interval (a, b), that $g(x) \neq 0$ and $g'(x) \neq 0$ for x in (a, b), and that*

$$\lim_{x \to a+} f(x) = 0 = \lim_{x \to a+} g(x). \tag{9}$$

Then

$$\lim_{x \to a+} \frac{f(x)}{g(x)} = \lim_{x \to a+} \frac{f'(x)}{g'(x)} \tag{10}$$

whenever the limit on the right side exists. (The limits here are limits from the right at a.)

Proof: The assumption (9) means that if we define $f(a) = 0 = g(a)$ then f and g are continuous on the interval $[a, b]$. Suppose that the limit on the right side in (10) exists and equals L. Given any $\varepsilon > 0$, choose $\delta > 0$ such that

$$\left| \frac{f'(c)}{g'(c)} - L \right| < \varepsilon \quad \text{if } a < c \leq a + \delta.$$

For any $x \in (a, a + \delta)$ we may apply the GMVT on the interval $[a, x]$ to conclude that

$$\left| \frac{f(x)}{g(x)} - L \right| < \varepsilon. \quad \square$$

Remark. It can be useful to know that the conclusion (10) is also true if f and g are differentiable on (a, ∞) and instead of (9) we have

$$\lim_{x \to \infty} f(x) = \infty = \lim_{x \to \infty} g(x), \tag{9'}$$

when the limit on the right in (10) exists. In fact, given $0 < \varepsilon < 1$, choose b so large that $|f'(c)/g'(c) - L| < \varepsilon$ for all $c > b$. Note that (9') implies that, for all sufficiently large x,

$$|f(b)| < \varepsilon|f(x)|, \qquad |g(b)| < \varepsilon|g(x)|.$$

Then

$$\left|\frac{f(x)}{g(x)} - \frac{f(x)-f(b)}{g(x)-g(b)}\right| = \left|\frac{f(x)g(b) - f(b)g(x)}{[g(x)-g(b)]g(x)}\right|$$

$$\leq 2\varepsilon \left|\frac{f(x)}{g(x)-g(b)}\right| \leq \frac{2\varepsilon}{1-\varepsilon}\left|\frac{f(x)-f(b)}{g(x)-g(b)}\right|.$$

Then again the GMVT leads to the result.

Exercises

1. Compute the following limits, with justification. (In (f), assume that $f''(x)$ is continuous.)

   (a) $\displaystyle\lim_{x \to 1} \frac{x \log x}{e^x - e}$.

   (b) $\displaystyle\lim_{x \to 0} \frac{\log(1+x) - x}{\sin(x^2)}$.

   (c) $\displaystyle\lim_{x \to \infty} x^{-a} \log x, \quad a > 0$.

   (d) $\displaystyle\lim_{x \to 0} x^a \log x, \quad a > 0$.

   (e) $\displaystyle\lim_{x \to 0} \frac{1 - \cos x}{x \sin x}$.

   (f) $\displaystyle\lim_{h \to 0} \frac{f(x+h) + f(x-h) - 2f(x)}{h^2}$.

2. Suppose that f is a real-valued function on $\mathbb{R}$ whose derivative exists at each point and is bounded. Prove that f is uniformly continuous.

3. (a) Suppose that $f : \mathbb{R} \to \mathbb{R}$ is continuous and $\lim_{|x| \to \infty} f(x) = 0$. Prove that f is uniformly continuous.
   (b) Find a bounded function $f : \mathbb{R} \to \mathbb{R}$ such that f is differentiable at every point and uniformly continuous, but f' is not bounded.

4. (a) Suppose $f : \mathbb{R} \to \mathbb{R}$ and $|f(x)| \leq x^2$, all x. Prove that f is differentiable at $x = 0$.
   (b) Find a function $f : \mathbb{R} \to \mathbb{R}$ that is differentiable at one point and not continuous at any other point.

5. Suppose that f is differentiable at each point of (a, b) and suppose that the derivative is never 0. Prove that f is either strictly increasing or strictly decreasing on the interval. (Notice that f' is *not* assumed to be continuous.)

6. Suppose that f is differentiable at each point of the interval $[a, b]$ and suppose that $f'(a) < c < f'(b)$. Prove that there is a point x in (a, b) such that $f'(x) = c$. (Notice that f' is not assumed to be continuous.)

7. Let $f(0) = 0$ and $f(x) = \sin(1/x)$ if $x \neq 0$. Let $g(x) = \int_0^x f$. Prove that g is differentiable at every point, but the derivative is not continuous at $x = 0$.

8B. Inverse Functions

The following theorem is basic.

Theorem 8.9: Intermediate Value Theorem. *If f is a continuous real-valued function on the interval $[a, b]$ and $f(a) \neq f(b)$, then for each real number c between $f(a)$ and $f(b)$ there is a point x in the interval (a, b) such that $f(x) = c$.*

Proof: Let $g(x) = f(x) - c$; we want a point where $g(x) = 0$. We construct a sequence of intervals as follows: Let $I_0 = [a, b]$. If g vanishes at the midpoint of this interval, then we may stop. Otherwise, g changes sign (has different signs at the two endpoints) on one of the two closed subintervals of length $(b - a)/2$ that cover I_0. Denote this subinterval by I_1 and continue. Either we reach in finitely many steps a midpoint at which $g = 0$ or we obtain a sequence of intervals $I_n = [a_n, b_n]$ of length $(b - a)/2^n$, on each of which g changes sign. The sequences $\{a_n\}$ and $\{b_n\}$ have a common limit x. By continuity and the change of sign condition, $g(x) = 0$. □

We use the term *interval*, without qualification, to mean any of the possibilities (a, b) with $-\infty \leq a < b \leq +\infty$, $(a, b]$ with $-\infty \leq a < b < +\infty$, $[a, b)$ with $-\infty < a < b \leq +\infty$, or $[a, b]$, $-\infty < a < b < +\infty$. Thus one possibility is $(-\infty, +\infty) = \mathbb{R}$.

Corollary 8.10. *If f is a continuous real-valued function on an interval I, then the image $f(I) = \{f(x) : x \in I\}$ is an interval.*

Theorem 8.11: Inverse functions. *If f is a continuous strictly monotone real-valued function on an interval I, then f has an inverse function g: g is defined on the interval $f(I)$ and*

$$g(f(x)) = x, \quad x \in I; \qquad f(g(y)) = y, \quad y \in f(I). \tag{11}$$

The function g is also continuous and strictly monotone.

If f is differentiable at an interior point x of I and $f'(x) \neq 0$, then g is differentiable at $f(x)$ and

$$g'(f(x)) = \frac{1}{f'(x)}. \tag{12}$$

Proof: Suppose that f is strictly increasing. It follows from this that for each y in the interval $f(I)$ there is a unique $x \in I$ such that $f(x) = y$; set $x = g(y)$. Then (11) is satisfied. The function g is strictly increasing. For continuity, suppose that $y = f(x) \in f(I)$ and $\varepsilon > 0$ are given. Suppose that x is an interior point; a slight change in the argument will deal with an endpoint. Choose x_1 and x_2 in I such that

$$x - \tfrac{1}{2}\varepsilon \leq x_1 < x < x_2 \leq x + \tfrac{1}{2}\varepsilon.$$

Set $y_j = f(x_j)$ and let δ be the smaller of $y_2 - y$ and $y - y_1$. Then $|y' - y| < \delta$ implies that y' is in the interval (y_1, y_2), so $g(y')$ is in the interval (x_1, x_2), so $|g(y') - g(y)| < \varepsilon$.

Finally, suppose that x is an interior point and $f'(x)$ exists and is positive. Each y' close to y is $f(x')$ for $x' = g(y')$ close to x, so

$$\frac{g(y') - g(y)}{y' - y} = \frac{x' - x}{f(x') - f(x)}$$

and the limit is (12). □

Remark. The function $f(x) = x^3$ is strictly increasing on $\mathbb{R}$ and differentiable at every point, but the derivative vanishes at $x = 0$ and the inverse function $g(y) = y^{1/3}$ is not differentiable at $y = f(0) = 0$.

As an example, consider the real exponential function from Section 5C:

$$e^x = E(x) = \sum_{n=0}^{\infty} \frac{x^n}{n!} = 1 + x + \frac{x^2}{2} + \frac{x^3}{6} + \frac{x^4}{24} + \frac{x^5}{120} + \cdots \quad (13)$$

Theorem 8.12. *The exponential function is a continuous strictly increasing function from $\mathbb{R}$ onto $(0, +\infty)$. The inverse function $\log y$ has derivative $1/y$ at the point $y > 0$.*

Proof: It follows from (13) that $e^x > 0$ for $x \geq 0$. But since $e^{-x}e^x = e^{x-x} = 1$, it follows that $e^{-x} = 1/e^x$ and the function also takes positive values on $(-\infty, 0)$. Differentiating (13), one sees that the derivative is e^x, and hence positive, so the function is continuous and strictly increasing. Next, (13) implies $e^x > 1 + x$ when $x > 0$, so $e^{-x} < 1/(1+x)$ and

$$\lim_{x \to +\infty} e^x = +\infty; \quad \lim_{x \to -\infty} e^x = 0.$$

Thus the image is $(0, +\infty)$. The inverse function $g(y) = \log y$ satisfies $g'(y) = 1/f'(g(y)) = 1/f(g(y)) = 1/y$. □

Exercises

(For some of the following exercises, integration is assumed.)

1. Define $f : (-\pi/2, \pi/2) \to \mathbb{R}$ by $f(0) = 0$ and
$$f(x) = \frac{x - \sin x}{1 - \cos x}, \quad \text{if } x \neq 0.$$
   (a) Show that f is continuous at $x = 0$.
   (b) Show that f is strictly increasing and the image of the interval is all of $\mathbb{R}$.

2. Prove that for every $a \geq 0$ there is a unique $b \geq 0$ such that
$$a = \int_0^b \frac{dx}{(1 + x^3)^{1/5}}$$

3. (a) Prove that every polynomial of odd degree, having real coefficients, has a real root.
   (b) Prove that every polynomial of even degree, having real coefficients, attains a maximum or a minimum.
   (c) Give another proof of (a).

4. Suppose that $a_0, a_1, a_2, \ldots, a_n$ are real numbers such that
$$a_0 + \frac{a_1}{2} + \frac{a_2}{3} + \cdots + \frac{a_n}{n+1} = 0.$$
   Prove that the polynomial $a_n x^n + a_{n-1} x^{n-1} + \cdots + a_1 x + a_0$ has a root in the interval $(0, 1)$.

5. Suppose that f is a continuous real-valued function on $\mathbb{R}$ and suppose that $f(x)$ is rational whenever x is irrational. Prove or disprove: f must be constant.

8C. Integral Calculus

Suppose that $I = [a, b]$ is a closed, bounded interval. Recall that a function f from I to $\mathbb{R}$ is said to be bounded if its image $f(I)$ is bounded. This means that there is a constant M such that $|f(x)| \leq M$, all $x \in I$. Suppose that f is bounded.

Definitions. A *partition* of the interval $[a, b]$ is a collection P of points $x_0, x_1, \ldots, x_n$ such that
$$a = x_0 < x_1 < x_2 < \cdots < x_n = b.$$

The *lower sum* $L(f, P)$ and *upper sum* $U(f, P)$ of f with respect to the partition P are the numbers

$$L(f, P) = \sum_{k=1}^n m_k (x_k - x_{k-1}), \quad m_k = \inf\{f(x) : x \in [x_{k-1}, x_k]\}; \quad (14)$$

$$U(f, P) = \sum_{k=1}^n M_k (x_k - x_{k-1}), \quad M_k = \sup\{f(x) : x \in [x_{k-1}, x_k]\}. \quad (15)$$

Note that $\sum_1^n (x_k - x_{k-1}) = b - a$. It follows immediately that, for each partition P,

$$m(b-a) \leq L(f, P) \leq U(f, P) \leq M(b-a),$$
$$m = \inf\{f(x) : x \in [a, b]\}; \quad M = \sup\{f(x) : x \in [a, b]\}. \tag{16}$$

Examples

1. If $f(x) = c$ for all x in $[a, b]$, then for every P the lower and upper sums are $L(f, P) = U(f, P) = c$.
2. Suppose that f is defined for $x \in [0, 1]$ by $f(x) = 1$ when $x \in \mathbb{Q}$ and $f(x) = 0$ when $x \notin \mathbb{Q}$. Then, for every partition P, $L(f, P) = 0$ and $U(f, P) = 1$.

Definition. A partition P' is said to be a *refinement* of the partition P if each point of P also belongs to P'.

Lemma 8.13. *If P and P' are partitions of $[a, b]$ and P' is a refinement of P, then*

$$L(f, P) \leq L(f, P') \leq U(f, P') \leq U(f, P). \tag{17}$$

Proof: Note that (16) is a special case of Lemma 8.13. In fact, P is a refinement of the trivial partition that consists of $y_0 = a$, $y_1 = b$. The general case of Lemma 8.13 follows by applying the analogue of (16) to each subinterval $[x_{k-1}, x_k]$ determined by P. □

Corollary 8.14. *If P and P' are any two partition of $[a, b]$, then*

$$L(f, P) \leq U(f, P'). \tag{18}$$

Proof: Choose a partition P'' that is a refinement both of P and of P'. Then

$$L(f, P) \leq L(f, P'') \leq U(f, P'') \leq U(f, P'). \quad \square \tag{19}$$

The inequality (16) shows that the upper and lower sums with respect to all possible partitions form a bounded set.

Definition. The *lower integral* $\underline{\int_a^b} f$ and the *upper integral* $\overline{\int_a^b} f$ of f on the interval $[a, b]$ are defined to be

$$\underline{\int_a^b} f = \sup_P \{L(f, P)\}; \quad \overline{\int_a^b} f = \inf_P \{U(f, P)\}, \tag{20}$$

where the supremum and infimum are taken over all partitions P of the interval $[a, b]$.

Proposition 8.15. $\underline{\int_a^b} f \leq \overline{\int_a^b} f$.

Proof: First, fix a partition P. The inequality (19) shows that $L(f, P)$ is a lower bound for the upper sums, so $L(f, P) \leq \overline{\int_a^b} f$. This is true for every partition P, so the upper integral is an upper bound for the lower sums. This proves $\underline{\int_a^b} f \leq \overline{\int_a^b} f$. □

Definitions. A bounded real-valued function f on an interval $[a, b]$ is said to be *integrable* on $[a, b]$ *in the sense of Riemann*, or *Riemann integrable* if $\underline{\int_a^b} f = \overline{\int_a^b} f$. If so, then the common value is denoted $\int_a^b f$ or by $\int_a^b f(x)\,dx$ (or by $\int_a^b f(t)\,dt$, etc.) and is called the *integral* of f on the interval $[a, b]$.

In the second example above, $\underline{\int_a^b} f = 0$ and $\overline{\int_a^b} f = 1$, so the function is not integrable.

Proposition 8.16. *A bounded real-valued function f on an interval $[a, b]$ is Riemann integrable if and only if for each $\varepsilon > 0$ there is a partition P such that*

$$U(f, P) - L(f, P) < \varepsilon. \tag{21}$$

Proof: Suppose that f is Riemann integrable on $[a, b]$. Given $\varepsilon > 0$, it follows from the definitions of the lower and upper integrals that there are partitions P' and P'' such that

$$\left(\int_a^b f\right) - \frac{\varepsilon}{2} < L(f, P'); \quad U(f, P'') < \left(\int_a^b f\right) + \frac{\varepsilon}{2}.$$

Let P be a refinement of P' and of P''. Then

$$\left(\int_a^b f\right) - \frac{\varepsilon}{2} < L(f, P') \leq L(f, P) \leq U(f, P) \leq U(f, P'') < \left(\int_a^b f\right) + \frac{\varepsilon}{2},$$

which implies that $U(f, P) - L(f, P) \leq \varepsilon$.

Conversely, (21) implies that $0 \leq \overline{\int_a^b} f - \underline{\int_a^b} f < \varepsilon$. If this is true for each $\varepsilon > 0$, then $\underline{\int_a^b} f = \overline{\int_a^b} f$. □

Theorem 8.17: Integrability of continuous functions. *If f is a continuous real-valued function on an interval $[a, b]$, it is Riemann integrable on $[a, b]$.*

Proof: Since the interval $[a, b]$ is compact, it follows that f is bounded and uniformly continuous. Given $\varepsilon > 0$, there is a $\delta > 0$ such that if x and y are points of $[a, b]$ at distance $< \delta$, then $|f(x) - f(y)| < \varepsilon/(b - a)$. This means that if the partition P is chosen so that each of its intervals $[x_{k-1}, x_k]$ has length $< \delta$, then the m_k and M_k of (14) and (15) differ by $< \varepsilon/(b - a)$. Therefore (21) is satisfied. □

The following algebraic properties follow fairly readily from the definitions and from Lemma 8.13 and Proposition 8.16.

Proposition 8.18: Algebraic properties of the integral. *Suppose that f and g are Riemann integrable functions on $[a, b]$ and suppose that c is real. Then the functions cf, $f \pm g$ and $|f|$ are Riemann integrable on $[a, b]$ and*

$$\int_a^b cf(x)\,dx = c \int_a^b f(x)\,dx; \tag{22}$$

$$\int_a^b \left[f(x) \pm g(x)\right] dx = \int_a^b f(x)\,dx \pm \int_a^b g(x)\,dx; \tag{23}$$

$$\left|\int_a^b f(x)\,dx\right| \leq \int_a^b |f(x)|\,dx. \tag{24}$$

Moreover,

$$\int_a^b f(x)\,dx \leq \int_a^b g(x)\,dx \quad \text{if } f(x) \leq g(x) \text{ for all } x \in [a, b] \tag{25}$$

The additivity of the integral suggests an appropriate extension to complex-valued functions.

Definition. A bounded complex-valued function f on an interval $[a, b]$ is Riemann integrable if the real and imaginary parts are Riemann integrable. If so, we set

$$\int_a^b f(x)\,dx = \int_a^b \operatorname{Re} f(x)\,dx + i \int_a^b \operatorname{Im} f(x)\,dx. \tag{26}$$

In particular, f is Riemann integrable if it is continuous. □

The next result shows a different kind of additivity of the integral. It is easily proved by using partitions that include the point b.

Proposition 8.19. *Suppose that $a < b < c$. A bounded function f on the interval $[a, c]$ is Riemann integrable on $[a, c]$ if and only if it is Riemann integrable on each* of the subintervals $[a, b]$ and $[b, c]$. If so, then

$$\int_a^c f(x)\,dx = \int_a^b f(x)\,dx + \int_b^c f(x)\,dx. \tag{27}$$

The identity (27) is very useful and can be made more useful by dropping the conditions $a < b < c$. To do so, we assume that f is Riemann integrable on $[a, b]$ and *define*

$$\int_c^c f(x)\,dx = 0, \quad c \in [a, b]; \qquad \int_b^a f(x)\,dx = -\int_a^b f(x)\,dx. \tag{28}$$

Then it can be checked, case by case, that (27) is valid for any triple of real numbers a, b, and c for which all three integrals are defined.

Theorem 8.20: Differentiation of the integral. *Suppose that f is a continuous real- or complex-valued valued function on the interval I. Suppose that a is a point of I and let F be defined by*

$$F(x) = \int_a^x f(t)\,dt, \quad x \in I. \tag{29}$$

Then F is differentiable at every interior point x of I and $F'(x) = f(x)$.

Proof: Fix an interior point x of I. For y close to x we use (27) to conclude that

$$\frac{F(y) - F(x)}{y - x} = \frac{1}{y - x}\left[\int_a^y f(t)\,dt - \int_a^x f(t)\,dt\right]$$

$$= \frac{1}{y - x}\int_x^y f(t)\,dt,$$

and it follows that

$$\frac{F(y) - F(x)}{y - x} - f(x) = \frac{1}{y - x}\int_x^y \left[f(t) - f(x)\right]dt. \tag{30}$$

Given any $\varepsilon > 0$, we can choose $\delta > 0$ such that $|y - x| < \delta$ implies that the integrand in (30) has modulus $< \varepsilon$ at each point. Then the modulus of the right side of (30) is $< \varepsilon$. □

Corollary 8.21: Fundamental Theorem of Calculus. *Suppose that f and G are continuous real- or complex-valued functions on $[a, b]$. Suppose that G is differentiable at each $x \in (a, b)$ and $G'(x) = f(x)$. Then*

$$\int_a^b f(x)\,dx = G(b) - G(a). \tag{31}$$

Proof: Let F be defined by (29). Then the difference $F - G$ has derivative 0, so $F - G$ is constant. Therefore $F(b) - G(b) = F(a) - G(a) = -G(a)$, or $F(b) = G(b) - G(a)$. □

Remark. Another way to state this result is that

$$f(x) = f(a) + \int_a^x f'(t)\,dt \quad \text{if } f' \text{ is continuous.} \tag{32}$$

As an example, we obtain the integral form of the natural (base e) logarithm:

$$\log x = \int_1^x \frac{dt}{t}, \quad x > 0. \tag{33}$$

Exercises

1. (a) Suppose that f is a continuous real-valued function on the interval $[a, b]$. Prove that there exists $x \in [a, b]$ such that $\int_a^b f = f(x)(b - a)$.
   (b) Give another proof.
2. Let $f(0) = 0$ and $f(x) = \sin(1/x)$ for x in the interval $(0, 1]$. Prove that f is integrable on the interval $[0, 1]$.
3. Suppose that f is continuous and nonnegative on the interval $[a, b]$, where $a < b$, and suppose that $\int_a^b f = 0$. Prove that $f \equiv 0$ on $[a, b]$.
4. Let $g(1/n) = 1$ for $n \in \mathbb{N}$ and $g(x) = 0$ otherwise. Prove that g is integrable on the interval $[0, 1]$.
5. Let $f(p/q) = 1/q$ if the fraction p/q is in lowest terms and $f(x) = 0$ for irrational x. Prove that f is continuous at x if and only if x is irrational or $x = 0$.
6. Prove that the function in Exercise 5 is not differentiable at any point.
7. Prove that the function in Exercise 5 is integrable on the interval $[0, 1]$.
8. Suppose that f is continuous, nonnegative, and nondecreasing at each point of $[0, \infty)$. Prove that

$$\int_0^x f(t)\,dt \leq xf(x), \quad \text{all } x \geq 0.$$

8D. Riemann Sums

The proof of Theorem 8.17 shows that if f is continuous on $[a, b]$, then, for a given $\varepsilon > 0$, *any* partition P all of whose intervals are small enough will give upper and lower sums within ε of $\int_a^b f$. In particular, one might work (conceptually, at least) with partitions into equal subintervals. For a given $n \in \mathbb{N}$, let P_n denote the partition into n equal subintervals. Thus

$$x_j = a + \frac{j}{n}(b - a), \quad j = 0, 1, 2, \ldots, n.$$

(Note that these points are different for different values of n as well as for different values of j; we ought to, but will not, use a notation like $x_j^{(n)}$.) A corresponding *Riemann sum* is

$$R_n(f) = \sum_{j=1}^{n} f(x_j^*)(x_j - x_{j-1}), \qquad x_j^* \in [x_{j-1}, x_j].$$

This depends on the choice of the points x_j^*, as well as on n and f.

Clearly

$$L(f, P_n) \leq R_n(f) \leq U(f, P_n).$$

It follows from this and from the proof of Theorem 8.17 that

$$\lim_{n \to \infty} R_n(f) = \int_a^b f$$

whenever f is continuous on $[a, b]$.

Remark. There is a potential trap here. Let us return to Example 2 in Section 8C, on the interval $[0, 1]$. Here, given n, the endpoints $x_j = j/n$ of the subintervals are rational, as are the midpoints. Therefore, if we look at the Riemann sums with the x_j^*'s at the endpoints or midpoints of the subintervals, then each $R_n f = 0$. In other words, to see the failure of integrability it is necessary to look at more general choices of the points x_j^*.

Exercises

1. Determine the limit

$$\lim_{n \to \infty} \sum_{k=1}^{n} \left(\frac{k}{n}\right)^2 \frac{1}{n}.$$

2. Prove that

$$\lim_{N \to \infty} \sum_{n=-N}^{N} \left(\frac{1}{N + in} + \frac{1}{N - in}\right) = 2 \int_{-1}^{1} \frac{dt}{1 + t^2}.$$

8E*. Two Versions of Taylor's Theorem

Suppose that I is an open interval and that f is a complex-valued function on I with the property that f and its successive derivatives $f', f'', \ldots f^{(n)}$ are continuous on I.

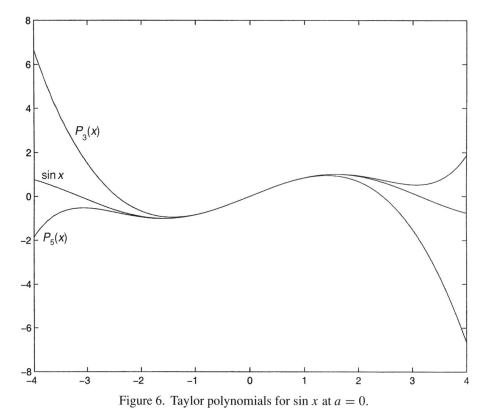

Figure 6. Taylor polynomials for sin x at $a = 0$.

At any given point $a \in I$, the n-th *Taylor polynomial* of f at a is the polynomial

$$T_{f,a}^{(n)}(x) = \sum_{k=0}^{n} \frac{f^{(k)}(a)}{k!}(x-a)^k$$

$$= f(a) + f'(a)(x-a) + \frac{f''(a)}{2}(x-a)^2 + \cdots + \frac{f^{(n)}(a)}{n!}(x-a)^n. \quad (34)$$

See Figure 6.

Proposition 8.22. *The Taylor polynomial $T_{f,a}^{(n)}$ is the unique polynomial p of degree at most n that has the property that*

$$p^{(k)}(a) = f^{(k)}(a), \qquad k = 0, 1, 2, \ldots, n. \quad (35)$$

Proof: One checks, by the simple expedient of differentiating, that $p = T_f^{(n)}$ has the property (35). Conversely, if p is a polynomial of degree at most n that satisfies (35) and we set $q(x) = p(x-a)$, then (35) determines $q^k(0), 0 \leq k \leq n$, which in turn determines the coefficients of q. Thus $p = T_f^{(n)}$. □

Theorem 8.23: Taylor's formula with remainder. *Suppose that f is a complex-valued function on an interval I with the property that f and its derivatives of order $\leq n$ are continuous, and suppose that a and x are points of I. Then there is a point c between a and x such that*

$$f(x) = f(a) + f'(a)(x-a) + \frac{f''(a)}{2}(x-a)^2 + \cdots + \frac{f^{(n-1)}(a)}{(n-1)!}(x-a)^{n-1}$$
$$+ \frac{f^{(n)}(c)}{n!}(x-a)^n. \tag{36}$$

Proof: For $n = 1$ this is just the Mean Value Theorem. Suppose that $n \geq 2$. Let $g(x) = f(x) - T_{f,a}^{(n-1)}(x)$ and $h(x) = (x-a)^n$. Then $g(a) = h(a) = 0$. By the Generalized Mean Value Theorem there is a point c_1 between a and x such that

$$\frac{g(x)}{h(x)} = \frac{g(x) - g(a)}{h(x) - h(a)} = \frac{g'(c_1)}{h'(c_1)}.$$

Now g' and h' both vanish at a, so there is a point c_2 between a and c_1 such that

$$\frac{g(x)}{f(x)} = \frac{g'(c_1)}{h'(c_1)} = \frac{g''(c_2)}{h''(c_2)}.$$

Continuing, we eventually reach a point $c = c_n$ such that

$$\frac{g(x)}{h(x)} = \frac{g^{(n)}(c)}{h^{(n)}(c)} = \frac{f^{(n)}(c)}{n!}.$$

Multiplication by $(x-a)^n$ converts this last equation to (36). □

Remark. Since continuity of the derivative is not required in the GMVT, we do not need to assume in Theorem 8.23 that $f^{(n)}$ is continuous, only that it exists at each point between a and x.

Just as the previous version of Taylor's formula starts with the Mean Value Theorem, the next version starts with the Fundamental Theorem of Calculus in the form (32). To extend (32) we can use integration by parts.

Theorem 8.24: Integration by parts. *Suppose that I is an interval and that f and g are complex-valued functions on I whose first derivatives are continuous. Suppose that a and b are points of I. Then*

$$\int_a^b f(x)g'(x)\,dx = f(x)g(x)\Big|_a^b - \int_a^b f'(x)g(x)\,dx. \tag{37}$$

Proof: This results from integrating Leibniz's rule in the form $fg' = (fg)' - f'g$. □

Fix a and x and set $h_k(t) = (x-t)^k$, so that $h_k(a) = 0$ for $k \geq 1$ and h_k is the derivative (with respect to t) of $-h_{k+1}/(k+1)$. Starting from (32) we obtain

$$\begin{aligned}
f(x) &= f(a) + \int_a^x f'(t)\,dt \\
&= f(a) - \int_a^x f'(t)\,h_1'(t)\,dt \\
&= f(a) + f'(a)h_1(a) + \int_a^x f''(t)h_1(t)\,dt \\
&= f(a) + f'(a)(x-a) - \frac{1}{2}\int_a^x f''(t)\,h_2'(t)\,dt \\
&= f(a) + f'(a)(x-a) + \frac{1}{2}f''(a)(x-a)^2 + \frac{1}{2}\int_a^x f'''(t)h_2(t)\,dt.
\end{aligned}$$

Continuing, one arrives at the following result.

Theorem 8.25: Taylor's formula with integral remainder. *Suppose that f is a complex-valued function on an interval I with the property that f and its derivatives of order $\leq n$ are continuous, and suppose that a and x are points of I. Then*

$$f(x) = f(a) + f'(a)(x-a) + \frac{f''(a)}{2}(x-a)^2 + \cdots + \frac{f^{(n-1)}(a)}{(n-1)!}(x-a)^{n-1}$$

$$+ \frac{1}{(n-1)!}\int_a^x f^{(n)}(t)(x-t)^{n-1}\,dt. \tag{38}$$

Exercises

1. (a) Prove that, for every positive integer n, $\lim_{x \to \infty} x^n e^{-x} = 0$.
   (b) Give another proof.
2. Let $f(x) = 0$ when $x \leq 0$ and $f(x) = \exp(-1/x)$ for $x > 0$. Prove that f and all its derivatives are continuous at $x = 0$. What are the Taylor polynomials for f at $a = 0$?
3. Suppose that f is a bounded real-valued function on $\mathbb{R}$, and suppose that its first and second derivatives are bounded and continuous. Prove that

$$\sup_{x \in \mathbb{R}} |f'(x)|^2 \leq 4 \sup_{x \in \mathbb{R}} |f(x)| \cdot \sup_{x \in \mathbb{R}} |f''(x)|.$$

Additional Exercises for Chapter 8

1. Suppose that f is real-valued and continuous on $[0, 1]$, and $f(0) = f(1) = 0$. Suppose that the second derivative f'' exists and is nonnegative at each point of $(0, 1)$. Prove that either f is constant on the interval or else $f(x) < 0$ for all $x \in (0, 1)$.

2. Suppose that f is real-valued on (a, b) and $f''(x) \geq 0$ for all $x \in (a, b)$. Prove that f is convex (see Exercise 3 of Section 7B).

3. Suppose that f is continuous from $[a, b]$ to $[c, d]$, and suppose that φ is continuous and convex on $[c, d]$. Prove *Jensen's inequality*:

$$\varphi\left(\frac{1}{b-a}\int_a^b f(x)\,dx\right) \leq \frac{1}{b-a}\int_a^b \varphi(f(x))\,dx.$$

4. Suppose that f is a continuous real-valued function defined on the unit square in $\mathbb{R}^2$: $\{(x, y) : 0 \leq x, y \leq 1\}$. Let

$$F(x) = \int_0^1 f(x, y)\,dy, \quad 0 \leq x \leq 1.$$

Prove that F is continuous.

5. Suppose that L is a real-valued function on $(0, \infty)$ with the properties:
   (i) $L(xy) = L(x) + L(y)$, all positive x and y;
   (ii) $L'(1)$ exists and $= 1$.
   Prove that $L(1) = 0$, and $L'(x) = 1/x$ for all $x > 0$.

6. Conversely, suppose that $L(x)$ is defined to be $\int_1^x dt/t$ for $x > 0$. Prove that L has the two properties of the preceding exercise.

7. Suppose that $\{f_n\}_1^\infty$ is a sequence in the space $C(I)$ of continuous functions on the interval $I = [a, b]$, with limit f. Prove that

$$\lim_{n\to\infty} \int_a^b f_n = \int_a^b f.$$

(In other words, for a uniformly convergent sequence, the limit of the integrals is the integral of the limit.)

8. Find an example of a sequence of real-valued continuous functions f_n on $[0, 1]$ such that $\lim_{n\to\infty} f_n(x) = 0$, all $x \in [0, 1]$, but $\int_0^1 f_n(x)\,dx = 1$, all n. Compare with the preceding exercise.

9. Suppose that h is a continuous real-valued function on $\mathbb{R}$. Define a function H on the metric space $C([a, b])$ by

$$H(f) = \int_a^b h(f(x))\,dx.$$

Prove that H is continuous.

10. Let $I = [0, 1]$ and let $L(f)$ be defined for f in $C(I)$ by

$$L(f) = \int_0^{\frac{1}{2}} f(x)\,dx - \int_{\frac{1}{2}}^1 f(x)\,dx.$$

This is a continuous function on $C(I)$. Let A be the bounded closed subset $A = \{f \in C(I) : \|f\| \leq 1\}$. Show that

$$\sup_{f \in A} L(f) = 1; \quad \text{but } L(f) < 1, \text{ all } f \in A.$$

11. Suppose that f is a continuous real-valued function on $[a, b]$, and suppose that for every nonnegative integer n
$$\int_a^b x^n f(x)\,dx = 0.$$
Prove that $f(x) = 0$, all $x \in [a, b]$.

12. Define $d_1(f, g)$ for f and g in $C([a, b])$ by
$$d_1(f, g) = \int_a^b |f(x) - g(x)|\,dx.$$
Show that d_1 is a metric on $C([a, b])$, but that $C([a, b])$ is not complete with respect to d_1.

9
Some Special Functions

So far we have dealt with few particular functions, apart from polynomials and the exponential functions. Here the basic trigonometric functions make their appearance, in a way that is particularly efficient for establishing their properties rigorously (but manages to hide the geometry until late in the game). One application is the Fundamental Theorem of Algebra; a second is Euler's product formula, which gives an evaluation of another of the series from Chapter 1.

9A. The Complex Exponential Function and Related Functions

We return to the function $E(z) = \sum_0^\infty z^n/n!$ of Section 5C. Based on the justification given there, we *define* $e^z = E(z)$. Now

$$e^{x+iy} = E(x+iy) = E(x) \cdot E(iy) = e^x \, e^{iy},$$

so to understand the complex exponential function we need to understand the real exponential function e^x and the function e^{iy}. We know that e^x is a strictly increasing function from $\mathbb{R}$ to $(0, +\infty)$ that is its own derivative, so in this section we examine $e^{ix}, x \in \mathbb{R}$. The power series is absolutely convergent for all x. We may group the real and imaginary parts to find that

$$e^{ix} = \sum_{n=0}^\infty \frac{i^n x^n}{n!} = 1 + ix - \frac{x^2}{2} - \frac{ix^3}{6} + \frac{x^4}{24} + \frac{ix^5}{120} - \frac{x^6}{720} - \frac{ix^7}{5040} + \cdots$$
$$= C(x) + i\,S(x),$$

where C and S are the functions from $\mathbb{R}$ into $\mathbb{R}$ defined by the power series

$$C(x) = \sum_0^\infty \frac{(-1)^n x^{2n}}{(2n)!} = 1 - \frac{x^2}{2} + \frac{x^4}{24} - \frac{x^6}{720} + \cdots ; \qquad (1)$$

$$S(x) = \sum_0^\infty \frac{(-1)^n x^{2n+1}}{(2n+1)!} = x - \frac{x^3}{6} + \frac{x^5}{120} - \frac{x^7}{5040} + \cdots . \qquad (2)$$

119

We will not assume that these functions are familiar in any way. Instead, we derive their properties from first principles. Notice that the complex conjugate of e^{ix} is e^{-ix} for real x. It follows that we can also write the real and imaginary parts in terms of the exponential function itself:

$$C(x) = \tfrac{1}{2}(e^{ix} + e^{-ix}); \qquad S(x) = \tfrac{1}{2i}(e^{ix} - e^{-ix}). \tag{3}$$

Proposition 9.1. *The functions C and S are differentiable and have the properties*

$$C'(x) = -S(x), \quad S'(x) = C(x), \quad \text{all } x \in \mathbb{R}; \qquad C(0) = 1, \quad S(0) = 0. \tag{4}$$
$$C(x)^2 + S(x)^2 = 1 \quad \text{all } x \in \mathbb{R}. \tag{5}$$

Proof: The identities (4) come from term-by-term differentiation of the power series (1) and (2), and from evaluation at $x = 0$.

One can deduce (5) in several ways; here are two. First, since $e^{ix} = C(x) + iS(x)$, we have

$$C(x)^2 + S(x)^2 = |e^{ix}|^2 = e^{ix}e^{-ix} = e^{ix-ix} = e^0 = 1.$$

Second, we derive (5) from (4). Set $g(x) = C(x)^2 + S(x)^2$. Then

$$g' = 2C\,C' + 2S\,S' = -2C\,S + 2S\,C = 0, \qquad g(0) = 1 + 0 = 1.$$

Therefore g is constant and $g(x) = 1$, all x. □

It is not surprising that one can deduce (5) directly from (4). In fact, the next result shows that the properties (4) completely determine the pair of functions C and S.

Proposition 9.2: Uniqueness of C and S. *If C_1 and S_1 are two real-valued differentiable functions with the properties $C_1' = -S_1$, $S_1' = C_1$, $C_1(0) = 1$, and $S_1(0) = 0$, then $C_1 = C$ and $S_1 = S$.*

Proof: Let h be the function $h = (C - C_1)^2 + (S - S_1)^2$. Then $h(0) = (1 - 1)^2 + 0 = 0$, so it is enough to show that h is constant. But

$$h' = 2(C - C_1)(-S + S_1) + 2(S - S_1)(C - C_1) = 0. \quad \square$$

Lemma 9.3. *There is a positive number x_0 such that*

$$\begin{aligned} C(x) > 0, \quad 0 \le x < x_0; & \qquad C(x_0) = 0; \\ S(x) > 0, \quad 0 < x \le x_0; & \qquad S(x_0) = 1. \end{aligned} \tag{6}$$

9A. The Complex Exponential Function and Related Functions

Proof: The series that defines the function C is an alternating series so long as x is in the interval $[0, 1]$, so on this interval $C(x) > 1 - x^2/2 \geq 1/2$. Suppose now that C is nonnegative on the interval $[1, a]$, where $a > 1$. Since $S' = C$, it follows that S is nondecreasing on this interval. The series that defines $S(1)$ is an alternating series, so $S(1) > 1 - 1/6 = 5/6$. Therefore $C'(x) = -S(x) \leq -5/6$ on $[1, a]$ and so

$$0 \leq C(a) = C(1) + \int_1^a C'(x)\, dx < 1 - 5(a-1)/6,$$

which implies $a < 11/5$. Therefore C becomes negative somewhere on the interval $[1, 11/5]$. By continuity and the Intermediate Value Theorem, there is a smallest number x_0 in this interval at which C vanishes. Thus $C = S'$ is positive on $[0, x_0)$, and it follows that S is strictly increasing on this interval and hence positive on $(0, x_0]$. Since $C^2 + S^2 = 1$ and $C(x_0) = 0$, it follows that $S(x_0) = 1$. □

We now *define* the number π to be $2x_0$; thus, $\pi/2$ is the smallest positive number at which the function C vanishes. We shall eventually relate π to the circle. (The proof of Lemma 9.3 shows that for π as just defined, $2 < \pi < 22/5$.)

Proposition 9.4. *The functions S and C have the properties, for all real x,*

$$C\left(x + \tfrac{1}{2}\pi\right) = -S(x), \qquad S\left(x + \tfrac{1}{2}\pi\right) = C(x); \tag{7}$$
$$C(x + \pi) = -C(x), \qquad S(x + \pi) = -S(x); \tag{8}$$
$$C(x + 2\pi) = C(x), \qquad S(x + 2\pi) = S(x). \tag{9}$$

Proof: Set $C_1(x) = S(x + \tfrac{1}{2}\pi)$ and $S_2(x) = -C(x + \tfrac{1}{2}\pi)$. These functions have the properties (4), so they coincide with S and C. This proves (7). The remaining identities are consequences of (7). For example,

$$C(x + \pi) = C\big((x + \tfrac{1}{2}\pi) + \tfrac{1}{2}\pi\big) = -S\big(x + \tfrac{1}{2}\pi\big) = -C(x). \quad \square$$

We can now demonstrate the geometric significance of these functions.

Theorem 9.5. *The function $F : [0, 2\pi] \to \mathbb{R}^2$, defined by $F(x) = (C(x), S(x))$, takes this interval onto the circle with center $(0, 0)$ and radius 1. As x increases, $F(x)$ moves in the direction of increasing argument and the length of the arc of the circle from $F(0)$ to $F(x)$ is x.*

In view of this result we set

$$\cos x = C(x), \qquad \sin x = S(x) \tag{10}$$

and note that these quantities have their usual geometric significance. It follows that one can link these trigonometric functions to the complex exponential function:

$$e^{ix} = \cos x + i \sin x; \qquad \cos x = \frac{e^{ix} + e^{-ix}}{2}; \qquad \sin x = \frac{e^{ix} - e^{-ix}}{2i}. \tag{11}$$

One consequence of this linkage is to relate the addition formula for the exponential to the addition formulas for sine and cosine:

$$\begin{aligned}\cos(x+y) + i\sin(x+y) &= e^{i(x+y)} = e^{ix} e^{iy} \\ &= (\cos x + i \sin x)(\cos y + i \sin y) \\ &= (\cos x \cos y - \sin x \sin y) + i(\cos x \sin y + \sin x \cos y).\end{aligned}$$

Equate the real and imaginary parts on left and right to obtain

$$\begin{aligned}\cos(x+y) &= \cos x \cos y - \sin x \sin y; \\ \sin(x+y) &= \cos x \sin y + \sin x \cos y.\end{aligned} \tag{12}$$

Proof of Theorem 9.5: Since $C^2 + S^2 = 1$, the function F maps into the unit circle. The Intermediate Value Theorem and Proposition 9.4 imply that this map hits every point and proceeds in the counter-clockwise direction. For example, if $a^2 + b^2 = 1$ and $a, b > 0$, then there is a unique $x \in (0, \frac{1}{2}\pi)$ such that $C(x) = a$ and then necessarily $S(x) = b$.

To prove the statement about arc length, we must first give a reasonable definition of arc length. Given $x \in [0, 2\pi]$ and $n \in \mathbb{N}$, set

$$x_{kn} = \frac{k}{n} x, \qquad k = 0, 1, 2, \ldots, n.$$

Thus $x_{kn} - x_{k-1,n} = x/n$. By continuity, as n increases, the adjacent points of $\{F(x_{kn}) : 0 \le k \le n\}$ become close on the circular arc joining $F(0)$ to $F(x)$. Thus it is reasonable to take the length of the arc to be

$$\lim_{n \to \infty} \sum_{k=1}^{n} |F(x_{kn}) - F(x_{k-1,n})|, \tag{13}$$

provided this limit exists, where $|\ |$ denotes the euclidean norm. Thus

$$|F(t) - F(s)|^2 = [C(t) - C(s)]^2 + [S(t) - S(s)]^2.$$

By the Mean Value Theorem there are points t', t'' between s and t such that

$$\begin{aligned}|F(t) - F(s)|^2 &= [C'(t')(t-s)]^2 + [S'(t'')(t-s)]^2 \\ &= [S(t')^2 + C(t'')^2](t-s)^2.\end{aligned}$$

Applying the Mean Value Theorem once again,
$$\left|C(t'')^2 - C(t')^2\right| \le 2\left|t'' - t'\right| \le 2|t - s|.$$

Since $S(t')^2 + C(t')^2 = 1$, we find that
$$\left||F(t) - F(s)|^2 - (t-s)^2\right| \le 2|t-s|^3,$$

so
$$\left||F(t) - F(s)| - |t-s|\right| \le \frac{2|t-s|^3}{|F(t) - F(s)| + |t-s|} \le 2(t-s)^2.$$

It follows from the preceding that
$$\left|\sum_{k=1}^{n} |F(x_{kn}) - F(x_{k-1,n})| - x\right| \le \frac{2x^2}{n},$$

and so the limit (13) is x. □

Exercises

1. Show that $e^{z_1} = e^{z_2}$ if and only if $z_1 - z_2 = 2n\pi i$ for some integer n.
2. Suppose that w is a nonzero complex number.
   (a) Find all complex z such that $e^z = w$.
   (b) Writing $z = x + iy$, show that e^x and y are polar coordinates of the point $(\operatorname{Re} w, \operatorname{Im} w) \in \mathbb{R}^2$.
3. Show that, for any complex $w \ne 0$ and any positive integer n, the equation $z^n = w$ has exactly n distinct complex solutions z, and find them explicitly.
4. The *hyperbolic functions* hyperbolic sine ("sinh") and hyperbolic cosine ("cosh") are defined by
$$\cosh x = \tfrac{1}{2}(e^x + e^{-x}), \qquad \sinh x = \tfrac{1}{2}(e^x - e^{-x}).$$

   (a) Prove that $\cosh^2 x - \sinh^2 x \equiv 1$.
   (b) Prove that $\cosh(x + y) = \cosh x \cosh y + \sinh x \sinh y$.
   (c) Prove that $\sinh(x + y) = \sinh x \cosh y + \cosh x \sinh y$.
5. Extend (11) to arbitrary complex z:
$$\cos z = \tfrac{1}{2}(e^{iz} + e^{-iz}), \qquad \sin z = \tfrac{1}{2i}(e^{iz} - e^{-iz}).$$

   (a) Determine the power series expansions of these functions.
   (b) Prove that $e^{iz} = \cos z + i \sin z$ for all $z \in \mathbb{C}$.
   (c) Prove that the addition formulas (12) extend to all complex values of the arguments.
   (d) Prove that $(\cos z)^2 + (\sin z)^2 = 1$, all $z \in \mathbb{C}$.
   (e) Give a second proof of (d).
   (f) Show that $\cos(ix) = \cosh x$ and $\sin ix = i \sinh x$, $x \in \mathbb{R}$.

6. For what real (or complex) values of x does the series $\sum_1^\infty 2^n \sin(3^{-n}x)$ converge?
7. True or false? For each $w \in \mathbb{C}$ there is $z \in \mathbb{C}$ such that $\cos z = w$. Find all solutions, when there are any.
8. Prove:

$$\sin\left(\tfrac{1}{2}x\right)[1 + 2\cos x + 2\cos 2x + 2\cos 3x + \cdots + 2\cos nx] = \sin\left(nx + \tfrac{1}{2}x\right).$$

9B*. The Fundamental Theorem of Algebra

Starting from $\mathbb{Q}$, one has to enlarge the field of numbers in order to have solutions even to simple equations like $x^2 - 2 = 0$. Even from $\mathbb{R}$ one has to go to $\mathbb{C}$ to have a solution to $x^2 + 1 = 0$. How does one know when to stop?

Theorem 9.6: Fundamental Theorem of Algebra. *A polynomial*

$$z^n + a_{n-1}z^n + \cdots + a_1 z + a_0, \qquad a_0, a_1, \ldots, a_{n-1} \in \mathbb{C} \tag{14}$$

of degree $n \geq 1$ has n complex roots $r_1, \ldots, r_n$ (counting multiplicity); it can be factored as

$$(z - r_1)(z - r_2)\ldots(z - r_n). \tag{15}$$

Proof: The key step is to show that the polynomial $p(z)$ defined by (14) has at least one root r. To see that this is the case, note that

$$\begin{aligned} |z| \geq R &= 1 + 2|a_0| + |a_1| + \cdots + |a_{n-1}| \Rightarrow \\ |p(z)| &\geq |z^n| - |a_{n-1}z^{n-1} + \cdots + a_0| \\ &\geq |z^n| - [|a_{n-1}| + \cdots + |a_0|]\,|z|^{n-1} \\ &\geq |z| - [|a_{n-1}| + \cdots + |a_0|] \geq |a_0| = |p(0)|. \end{aligned} \tag{16}$$

The continuous function $|p|$, restricted to the closed disk $\{z : |z| \leq R\}$, attains a minimum at a point $z = r$ and (16) implies that the minimum is a global minimum: $|p(r)| \leq |p(z)|$, all $z \in \mathbb{C}$. We shall see that this is only possible if $p(r) = 0$. In fact, suppose that s is such that $p(s) \neq 0$. The polynomial p may be written in the form

$$p(z) = b_0 + b_k(z - s)^k + b_{k+1}(z - s)^{k+1} + \cdots + (z - s)^n, \qquad b_k \neq 0,$$

where $b_0 = p(s) \neq 0$. Suppose that c is chosen with $|c| = 1$ and let $z - r = \varepsilon c$, where $0 < \varepsilon < 1$. Then

$$p(s + \varepsilon c) = b_0 + (b_k c^k)\varepsilon^k + h(\varepsilon, c)\varepsilon^{k+1}, \qquad |h(\varepsilon, c)| \leq K.$$

We can choose c so that $c^k = -\rho b_0/b_k$, where $\rho > 0$ (see Exercise 3 of Section 9A). Then

$$p(r + \varepsilon c) = b_0(1 - \varepsilon^k \rho) + h(\varepsilon, c)\varepsilon^{k+1}, \tag{17}$$

and for small positive ε the expression on the right side of (17) will have modulus smaller than $|b_0|$: It will be closer to the origin by a small multiple of ε^k. This proves that at a point r where $p(z)$ has minimum modulus, necessarily $p(r) = 0$. Thus p has at least one root.

Given a root r_1 for p, let $a_n = 1$ and note that

$$p(z) = p(z) - p(r_1) = \sum_{k=1}^{n} a_k\left(z^k - r_1^k\right) = (z - r_1)p_1(z),$$

where p_1 is the polynomial of degree $n - 1$:

$$p_1(z) = \sum_{k=1}^{n} a_k\left[z^{k-1} + z^{k-2}r_1 + \cdots + r_1^{k-1}\right].$$

The proof is completed by finding a root r_2 for p_1 and continuing. $\square$

9C*. Infinite Products and Euler's Formula for Sine

Our aim in this section is to prove Euler's formula for the sine function:

$$\sin x = x \prod_{n=1}^{\infty}\left(1 - \frac{x^2}{\pi^2 n^2}\right) \tag{18}$$

$$= x\left(1 - \frac{x^2}{\pi^2}\right)\left(1 - \frac{x^2}{4\pi^2}\right)\left(1 - \frac{x^2}{9\pi^2}\right)\cdots.$$

There is at least some plausibility to (18) because it appears that the right-hand side vanishes if and only if x is an integer multiple of π. If we accept that (18) has a precise meaning *and* that the product can be expanded out, then

$$x - \left(\sum_{n=1}^{\infty} \frac{1}{\pi^2 n^2}\right)x^3 + \cdots = \sin x = x - \frac{x^3}{6} + \cdots. \tag{19}$$

Comparing the coefficients of x^3,

$$\sum_{n=1}^{\infty} \frac{1}{n^2} = \frac{\pi^2}{6}. \tag{20}$$

(Several confirmations of this result are given in the exercises for Chapter 13.) There are three main steps to justifying (18). One is to give a meaning to an infinite product. Consider a formal product

$$\prod_{n=1}^{\infty} b_n = b_1 b_2 b_3 \cdots . \tag{21}$$

In analogy with infinite series, one might take this to mean the limit of the partial products

$$p_n = \prod_{k=1}^{n} b_k = b_1 b_2 \cdots \cdots b_n. \tag{22}$$

Notice, however, that if some $b_m = 0$, then $p_n = 0$ for all $n \geq m$, no matter what the other factors are. Moreover, if $|b_n| \leq r < 1$ for all large enough n, then the p_n's converge to 0, but this is not a particularly interesting result. For these reasons one defines a formal product to be *convergent* if and only if for each $\varepsilon > 0$ there is an index N such that

$$|b_m b_{m+1} \cdots \cdot b_n - 1| < \varepsilon \qquad \text{if } n \geq m \geq N. \tag{23}$$

This condition implies that the partial products (22) converge and that the limit is zero if and only if some factor $b_m = 0$. We also write the limit as $\prod_{1}^{\infty} b_n$. In particular, (23) implies that $\lim_{n \to \infty} b_n = 1$. For this reason the factors b_n are commonly written in the form $b_n = 1 + a_n$. Note that

$$|(1+a_m)(1+a_{m+1}) \cdots \cdot (1+a_n) - 1|$$
$$\leq (1+|a_m|)(1+|a_m|) \cdots \cdot (1+|a_n|) - 1. \tag{24}$$

The product $\prod_{1}^{\infty}(1+a_n)$ is said to be *absolutely convergent* if $\prod_{1}^{\infty}(1+|a_n|)$ is convergent. It follows from (24) that absolute convergence implies convergence. Now the product $\prod_{m}^{n}(1+|a_m|)$ is close to 1 if and only if its logarithm is close to zero. Since

$$\frac{a}{1+a} \leq \int_0^a \frac{dt}{1+t} \leq a, \qquad \text{if } a \geq 0,$$

it follows that

$$\frac{a}{2} \leq \log(1+a) \leq a \qquad \text{if } 0 \leq a \leq 1 \tag{25}$$

Taking all this into consideration, we obtain the following criterion.

Proposition 9.7. *The product $\prod_{1}^{\infty}(1+a_n)$ is absolutely convergent if and only if the series $\sum_{1}^{\infty} a_n$ is absolutely convergent.*

9C*. Infinite Products and Euler's Formula for Sine

In particular, the infinite product in (18) is absolutely convergent for every real x (in fact, for every complex x as well). Moreover, one can use the estimates above to justify the expansion of this product as a power series in x. It remains to prove that the value of the product is $\sin x$. The next step is to represent the complex exponential function as a limit of products.

Proposition 9.8. *For any complex z,*

$$\lim_{n\to\infty} \left(1 + \frac{z}{n}\right)^n = e^z. \tag{26}$$

Proof: For real z this was done in Section 3G. In general,

$$e^z - \left(1 + \frac{z}{n}\right)^n = \sum_{k=0}^{\infty} \frac{z^k}{k!} - \sum_{k=0}^{n} \binom{n}{k}\frac{z^k}{n^k} = \sum_{k=0}^{n} c_{nk} \frac{z^k}{k!} + \sum_{k=n+1}^{\infty} \frac{z^k}{k!}, \tag{27}$$

where

$$0 \le c_{nk} = 1 - \frac{n!}{(n-k)!\, n^k} = 1 - \prod_{j=0}^{k-1} \frac{n-j}{n} \le 1.$$

Note that for each fixed k, $\lim_{n\to\infty} c_{nk} = 0$. Given z and given $\varepsilon > 0$, we may choose m so large that $\sum_{m+1}^{\infty} |z|^k/k! < \varepsilon$. Then as $n \to \infty$ the sum of the first m terms in the extreme right-hand side of (27) converges to 0 while the sum of the remaining terms has modulus $< \varepsilon$ for every n. Therefore (27) has limit 0. □

Because of (26),

$$\sin x = \frac{e^{ix} - e^{-ix}}{2i} = \lim_{n\to\infty} P_n(x),$$

where P_n is the polynomial

$$P_n(x) = \frac{1}{2i}\left(1 + \frac{ix}{n}\right)^n - \frac{1}{2i}\left(1 - \frac{ix}{n}\right)^n. \tag{28}$$

(See Figure 7.) This polynomial vanishes for $x = 0$ and also when $[(1 - ix/n)/(1 + ix/n)]^n = 1$, that is, when

$$x = in\frac{w-1}{w+1}, \qquad w^n = 1. \tag{29}$$

Suppose that $n = 2m$ is even. Then P_n has degree $n - 1 = 2m - 1$ and the roots of 1 that give the $2m - 2$ nonzero roots of P_n are $w = e^{ik\pi/m}$,

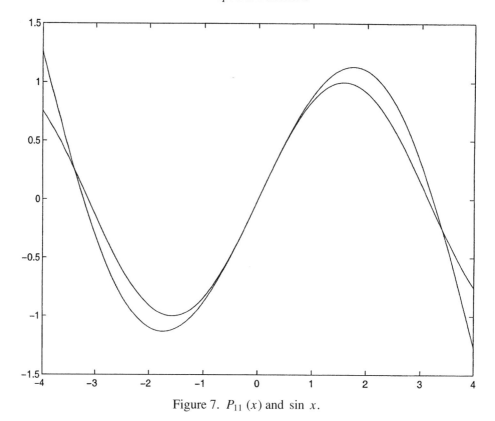

Figure 7. $P_{11}(x)$ and $\sin x$.

$k = \pm 1, \ldots k = \pm(m-1)$. The associated roots x are

$$x_{k,n} = i n \frac{e^{-ik\pi/m} - 1}{e^{-ik\pi/m} + 1} = i n \frac{e^{-ik\pi/2m} - e^{ik\pi/2m}}{e^{-ik\pi/2m} + e^{ik\pi/2m}}$$
$$= n \tan(k\pi/n) \qquad k = \pm 1, \pm 2, \ldots \pm (m-1).$$

Note that $P_n(x)/x = 1$ at $x = 0$. Consequently, the polynomial $P_n = P_{2m}$ can be factored as

$$P_{2m}(x) = x \prod_{k=1}^{m-1} \left(1 - \frac{x^2}{[2m \tan(k\pi/2m)]^2}\right). \tag{30}$$

Thus we have proved that

$$\sin x = \lim_{m \to \infty} x \prod_{k=1}^{m-1} \left(1 - \frac{x^2}{[2m \tan(k\pi/2m)]^2}\right). \tag{31}$$

Note that for each fixed k, $\lim_{n \to \infty} n \tan(k\pi/n) = k\pi$, so *each factor* in (31) converges to a factor of the product (18). We leave it to the reader to complete the proof by getting an appropriate estimate for the tail of the product expansion.

9C*. Infinite Products and Euler's Formula for Sine

Exercises

1. Carry the expansion (19) out further and show that $\sum_1^\infty 1/n^4 = \pi^4/90$.
2. Fill in the details of the proof of the Euler product formula: Prove that, for any fixed x and any $\varepsilon > 0$, there is an N such that

$$\left| \prod_N^{m-1} \left(1 - \frac{x^2}{[2m \tan(k\pi/2m)]^2} \right) - 1 \right| < \varepsilon$$

for every $m > N$. Conclude that, for large m, $P_{2m}(x)$ is close both to $\sin x$ and to the Euler product.

3. The *Gamma function* is the function defined for $s > 0$ by the integral

$$\Gamma(s) = \int_0^\infty e^{-x} x^s \frac{dx}{x}.$$

It is useful in many computations.
   (a) Show that $\Gamma(s+1) = s\, \Gamma(s)$.
   (b) Show that for positive integers n, $\Gamma(n) = (n-1)!$.
   (c) Prove the identity

$$s^{-a} = \frac{1}{\Gamma(a)} \int_0^\infty e^{-sx} x^a \frac{dx}{x}, \qquad s > 0,\ a > 0.$$

4. Prove that Γ is strictly decreasing on the interval $(0, 1)$ and strictly increasing on the interval $(1, \infty)$.
5. (a) Show that

$$\Gamma\left(\tfrac{1}{2}\right) = \int_{-\infty}^\infty e^{-x^2} dx.$$

   (b) Evaluate $\Gamma(1/2)$.

6. The *Beta function* is the function defined for $a > 0,\ b > 0$ by

$$\beta(a, b) = \int_0^1 t^{a-1} (1-t)^{b-1}\, dt.$$

It comes up in various calculations.
   (a) Show that

$$\int_0^x y^{a-1}(x-y)^{b-1}\, dy = \beta(a, b)\, x^{a+b-1}.$$

   (b) Use (a) and a change of variables in the double integral to prove that

$$\beta(a, b)\, \Gamma(a+b) = \Gamma(a)\, \Gamma(b).$$

7. Evaluate the following integral in terms of Gamma functions:

$$\int_0^\infty s^{a-1}(1+s)^{-1-b}\, ds, \qquad a > 0,\ b > 0.$$

8. The *hypergeometric function* is the function F defined for complex a, b, c, and z by the series

$$F(a,b,c;x) = \sum_{n=0}^{\infty} \frac{(a)_n (b)_n}{(c)_n n!} z^n,$$

where

$$(p)_0 = 1, \quad (p)_n = p(p+1)(p+2)\cdots(p+n-1), \quad n = 1, 2, 3, \ldots,$$

and it is assumed that c is not a negative integer.

(a) Prove that the radius of convergence of this series is at least 1.
(b) Under what conditions on a, b, c is the radius of convergence larger than 1?
(c) Verify that

$$(a)_n = \frac{\Gamma(a+n)}{\Gamma(a)}.$$

9. Prove that for $a > 0$ and $|z| < 1$, $f(a,b,b,z) = (1-z)^{-a}$.

10. Use Exercises 6(b), 8(c), and 9 to prove the identity

$$F(a,b,c;z) = \frac{1}{\beta(b, c-b)} \int_0^1 t^{a-1}(1-t)^{c-b}(1-tz)^{-a}\, dt$$

for $a > 0, c > b > 0, |z| < 1$.

10
Lebesgue Measure on the Line

The notions of length, area, and volume are more subtle than one might think, if one tries to extend them from simple sets to more general sets in the line, plane, or 3-space. For various reasons it is important to find such extensions, however, especially in connection with integration. In this chapter we introduce the basic construction and properties for subsets of the line.

10A. Introduction

The formulation of the integral in Chapter 8, due to Riemann, Darboux, and others, is adequate for many purposes. It assigns a number – the (definite) integral – to each bounded, piecewise continuous function on a bounded real interval. The procedure can be rephrased in the following way: Approximate the given function from above and below by *step functions*, that is, functions that are constant on each of finitely many subintervals whose union is the original interval. These approximating functions have integrals that should approach the desired value.

We have already introduced the standard example for which this procedure breaks down, the function

$$f : [0, 1] \to \mathbb{R}, \qquad f(x) = \begin{cases} 1 & \text{if } x \text{ is rational,} \\ 0 & \text{if } x \text{ is irrational.} \end{cases}$$

Approximation from above gives numbers ≥ 1, and from below it gives numbers ≤ 0.

This example may seem artificial, but consider it in the following way. If the function g takes only the values 0 and 1, then its integral should be the *total length* of the set $A = \{x : g(x) = 1\}$, since it represents the area of a figure of height 1 and base A. Therefore, trying to integrate this function f amounts to trying to determine the total length of the set consisting of all rationals in the interval $[0, 1]$, a geometric question that is not a priori meaningless. If there is a reasonable notion of total length, we can also ask whether the sum of the total length of $\mathbb{Q} \cap [0, 1]$ plus the total length of $\mathbb{Q}^c \cap [0, 1]$ is 1, the length of $[0, 1]$; as usual, $\mathbb{Q}$ denotes the rationals and $\mathbb{Q}^c$ denotes the complementary set (in $\mathbb{R}$), that is, the irrationals.

It is important to be aware that *reasonable-sounding geometric questions of this kind do not necessarily have answers*. For example, consider the question of *volume* in 3-space $\mathbb{R}^3$. A precise formulation of the question is, can we assign to each subset A of $\mathbb{R}^3$ a number, denoted $\text{vol}(A)$, that has the properties one would expect of volume:

(i) For each A, $0 \leq \text{vol}(A) \leq \infty$.
(ii) If A and B are disjoint, then $\text{vol}(A \cup B) = \text{vol}(A) + \text{vol}(B)$.
(iii) If A and B are congruent, then $\text{vol}(A) = \text{vol}(B)$.
(iv) If A is an open ball $\{x \in \mathbb{R}^3 : ||x - a|| < r\}$, where a is in $\mathbb{R}^3$ and $r > 0$, then $0 < \text{vol}(A) < \infty$.

The answer is no, it is *not* possible to accomplish this. The impossibility is shown by the following paradoxical result. We denote the congruence of sets A and B by $A \cong B$, meaning that there is a combination of translations and rotations of $\mathbb{R}^3$ that takes A onto B.

Theorem of Banach and Tarski. *Let A, B and C be pairwise disjoint closed balls of radius 1 in $\mathbb{R}^3$. There are pairwise disjoint sets $A_1, A_2, \ldots, A_7, B_1, B_2, B_3, C_1, C_2, C_3, C_4$ such that*

$$A = A_1 \cup A_2 \cup \cdots \cup A_7, \quad B = B_1 \cup B_2 \cup B_3, \quad C = C_1 \cup C_2 \cup C_3 \cup C_4;$$
$$A_1 \cong B_1, \quad A_2 \cong B_2, \quad A_3 \cong B_3, \quad A_4 \cong C_1, \quad A_5 \cong C_2, \quad A_6 \cong C_3, \quad A_7 \cong C_4.$$

In other words, we can disassemble *one* ball of radius 1 into seven pieces, move the pieces around by rotations and translations, and assemble them into *two* balls of radius 1. A proof is sketched in the Appendix. (In fact, it is possible to accomplish the same result with as few as *five* pieces, but no fewer. For more detail, see S. Wagon, The Banach-Tarski Paradox, Encyclopedia of Mathematics, vol. 24, Cambridge University Press, 1985.)

Exercise

1. Show that the Banach-Tarski theorem implies the impossibility of assigning volume to arbitrary subsets of $\mathbb{R}^3$ in such a way that properties (i)–(iv) are valid.

10B. Outer Measure

If I is an interval with endpoints a, b, $a \leq b$, the *length* of I is $b - a$. This also makes sense for the empty set $\emptyset = (a, a)$ and for infinite intervals:

$$|I| = b - a \quad \text{if } I = (a, b), [a, b), (a, b], \text{ or } [a, b].$$

We use the conventions $\infty - a = \infty$, and so on, from Section 3F.

Suppose that A is a subset of $\mathbb{R}$ that is *covered* by intervals $I_1, I_2, \ldots, I_N$, that is, $A \subseteq \bigcup_{n=1}^{N} I_n$. Then the "total length" of A should be at most $\sum_{n=1}^{N} |I_n|$. One might even take the greatest lower bound of all such numbers as the *definition* of the total length of A. However, this procedure has undesirable features. The set of rationals in [0, 1] and the set of irrationals in [0, 1] would each be assigned total length 1, as would the interval itself. Lebesgue took the important step of allowing *countable* covers.

Definition. If A is a subset of $\mathbb{R}$, the *outer measure* of A, denoted $m^*(A)$, is the infimum of all the numbers

$$\sum_{k=1}^{\infty} |I_k|,$$

which can be obtained by choosing a sequence of open intervals $\{I_k\}$ that covers A:

$$A \subseteq \bigcup_{k=1}^{\infty} I_k.$$

Here some of the intervals I_k (or all of them, if $A = \emptyset$) may be empty. In particular, $m^*(\emptyset) = 0$.

The following settles the question of the total length of the set of rationals in the unit interval or of all of $\mathbb{Q}$.

Example 1. Suppose that A has at most countably many points. Then $m^*(A) = 0$. In fact, let $(x_k)_{k=1}^{\infty}$ be an enumeration of the points of A. Given $\varepsilon > 0$, let I_k be an open interval of length $\varepsilon/2^k$ that contains the point x_k. The collection $\{I_k\}$ covers A, and $\sum |I_k| = \sum \varepsilon/2^k = \varepsilon$, so $m^*(A) \leq \varepsilon$.

In view of this example, it is fair to ask whether *every* set can be covered in some very clever way so as to show that its outer measure is zero. This is not the case.

Example 2. If A is an interval, then $m^*(A) = |A|$. To see this, suppose that A is a *bounded, closed* interval; the remaining cases can be deduced from this one and are left as an exercise. Suppose that $A = [a, b]$. Given $\varepsilon > 0$, we can choose

$I_1 = (a - \varepsilon/2, b + \varepsilon/2)$ and $I_k = \emptyset$ for $k > 1$ to see that $m^*(A) \leq b - a + \varepsilon$. Thus $m^*(A) \leq b - a = |A|$.

The converse is trickier. Suppose that $(I_k)_{k=1}^\infty$ is a cover of A by open intervals. Since A is compact, there is a finite subcover. Renumbering, assume that $I_1, I_2, \ldots, I_N$ cover A. Renumbering again and again as we go along, we may assume that the left endpoint a is in I_1 and either the right endpoint b is in I_1 or else the right endpoint of I_1 is less than b and belongs to I_2. Continuing to renumber, we eventually have points

$$a = a_1 < a_2 < \cdots < a_{n+1} = b, \qquad (a_j, a_{j+1}) \subseteq I_j.$$

Then

$$|A| = b - a = a_{n+1} - a_1 = \sum_{j=1}^n (a_{j+1} - a_j) \leq \sum_{j=1}^n |I_j| \leq \sum_{j=1}^\infty |I_j|,$$

so $m^*(A) \geq |A|$.

Example 3. The standard Cantor set C of Section 6E has $m^*(C) = 0$. In fact, $C = \bigcap_{n=0}^\infty C_n$, where $C_0 = [0, 1]$, $C_1 = [0, \frac{1}{3}] \cup [\frac{2}{3}, 1]$, and in general C_{n+1} consists of the intervals of C_n with their open middle thirds removed. Thus C_n is the union of 2^n closed intervals of total length $(\frac{2}{3})^n$. Since $C \subset C_n$, it follows that $m^*(C) \leq (\frac{2}{3})^n \to 0$.

Example 4. Suppose that A is a subset of $\mathbb{R}$ and h is real. The *translate* of A by h is the set

$$A + h = \{x + h : x \in A\}.$$

If I is an interval, then so is $I + h$, and the length is unchanged. If the intervals cover A, then their translates by h cover $A + h$. Therefore,

$$m^*(A + h) = m^*(A).$$

Here is a summary of the basic properties of outer measure. Some are obvious from the definition and some have been established in the examples, so only the third property needs to be discussed.

Properties of outer measure

(i) $\qquad 0 \leq m^*(A) \leq +\infty.$

(ii) $\qquad A \subset B$ implies $m^*(A) \leq m^*(B)$.

(iii) $\qquad A \subset \bigcup_{n=1}^\infty A_n$ implies $m^*(A) \leq \sum_{n=1}^\infty m^*(A_n)$.

(iv) $\quad m^*(A) = |A|$ if A is an interval.
(v) $\quad m^*(A + h) = m^*(A)$.

To establish property (iii), assume that $\varepsilon > 0$ is given. Choose a cover $(I_{n,k})_{k=1}^{\infty}$ of A_n such that

$$\sum_{k=1}^{\infty} |I_{n,k}| \leq m^*(A_n) + \frac{\varepsilon}{2^n}.$$

Then $(I_{n,k})_{n,k=1}^{\infty}$ is a countable cover of A of total length $\leq [\sum m^*(A_n)] + \varepsilon$.

There is one key property that one would *like* to have that is not on this list, and indeed is not necessarily true. Suppose that A and B are *disjoint* subsets of $\mathbb{R}$. We would like to have

$$m^*(A \cup B) = m^*(A) + m^*(B),$$

but this is not always the case. Rather than try to remedy this by modifying m^*, we restrict the class of subsets that we work with.

Exercises

1. Show that using closed intervals rather than open intervals in the definition of outer measure would not change the evaluation $m^*(A)$.
2. Using the result for closed, bounded intervals, prove that for *every* interval I, $m^*(I) = |I|$.
3. Show that for any two sets A and B with union $[0, 1]$, the outer measures satisfy $m^*(A) \geq 1 - m^*(B)$.
4. In Example 2 we saw that if $I_1, \ldots, I_n$ are open intervals whose union contains a given bounded closed interval A, then $|A| \leq \sum_{k=0}^{n} |I_k|$. Give another proof of this fact, by induction on n.
5. Suppose that I_1 and I_2 are disjoint open intervals, and suppose that A_1 is a subset of I_1 and A_2 is a subset of I_2. Prove, directly from the definition, that $m^*(A_1 \cup A_2) = m^*(A_1) + m^*(A_2)$.
6. Suppose that A is an open subset of $\mathbb{R}$. Prove that A is a countable union of disjoint open intervals (some or all of which may be empty).
7. Prove that if the open set A is the union of a sequence of disjoint open intervals $I_1, I_2, \ldots$, then $m^*(A) = \sum_{n=1}^{\infty} |I_n|$.
8. Prove that for any subset B of $\mathbb{R}$, $m^*(B) = \inf\{m^*(A) : A \text{ open}, A \supset B\}$.
9. The Cantor set, Example 3 of Section 10B, is a closed subset of $[0, 1]$ that is *nowhere dense*, that is, it contains no nonempty open intervals. For the standard Cantor set, $m^*(C) = 0$. Show that there is a *fat* Cantor set, a closed nowhere dense subset $A \subset \mathbb{R}$ of the interval $[0, 1]$, such that $m^*(A) > 0$. In fact, show that for each $\varepsilon > 0$ there is such a set with outer measure $m^*(A) > 1 - \varepsilon$.

10. Let $A = \{x \in [0, 1] : \text{no fives occur in the decimal expansion of } x\}$. Find $m^*(A)$.
11. Suppose that A is a bounded set and $m^*(A \cap I) \leq \frac{1}{2}m^*(I)$ for every interval I. Prove that $m^*(A) = 0$. What if A has the indicated property but is not bounded?

10C. Measurable Sets

Suppose that A and E are two subsets of $\mathbb{R}$. Let A^c denote the complement in $\mathbb{R}$. Then E is the union of the two disjoint pieces $E \cap A$ and $E \cap A^c$, but its outer measure is not necessarily the sum of the outer measure of the pieces. It turns out to be useful to select precisely those subsets A that *do* split every set E additively.

Definition. A subset A of $\mathbb{R}$ is *measurable* if, for every subset E of $\mathbb{R}$,

$$m^*(E) = m^*(E \cap A) + m^*(E \cap A^c). \tag{1}$$

Notice that, from property (iii) above, we always have

$$m^*(E) \leq m^*(E \cap A) + m^*(E \cap A^c).$$

Therefore, to prove that a given set A is measurable, it is sufficient to prove

$$m^*(E \cap A) + m^*(E \cap A^c) \leq m^*(E), \qquad \text{for all } E \subset \mathbb{R}. \tag{2}$$

Examples. The empty set and $\mathbb{R}$ are measurable. (Why?) If A is measurable, so is its complement. (Why?) If $m^*(A) = 0$, then A is measurable. (Why?)

We denote the collection of all measurable subsets of $\mathbb{R}$ by $\mathcal{M}$. In this section we prove one concrete and several abstract results on measurability of sets, beginning with the concrete result.

Proposition 10.1. *Every interval is measurable.*

Proof: Suppose that A is an interval and E is any subset of $\mathbb{R}$. Suppose that $(I_k)_1^\infty$ is a cover of E by open intervals and suppose that $\varepsilon > 0$ is given. Now $A \cap I_k$ is an interval and $A^c \cap I_k$ is either an interval or a union of two intervals, so it is easy to see that there are open intervals I_k', I_k'', and I_k''' such that

$$A \cap I_k \subset I_k', \qquad A^c \cap I_k \subset I_k'' \cup I_k''', \qquad |I_k'| + |I_k''| + |I_k'''| \leq |I_k| + \varepsilon/2^k.$$

Therefore $A \cap E \subset \bigcup_k I_k'$, $A^c \cap E \subset \bigcup_k (I_k'' \cup I_k''')$, and

$$\sum_{k=1}^\infty \left(|I_k| + \frac{\varepsilon}{2^k}\right) \geq \sum_{k=0}^\infty |I_k'| + \sum_{k=1}^\infty \left(|I_k''| + |I_k'''|\right) \geq m^*(A \cap E) + m^*(A^c \cap E).$$

Since this is true for every E and every $\varepsilon > 0$, we obtain (2). □

The abstract results mentioned above are the following five propositions. The first was already noted.

Proposition 10.2. *If A is measurable, then so is A^c.*

Proposition 10.3. *If A and B are measurable, then so are $A \cup B$ and $A \cap B$.*

Proof: Suppose that E is a subset of $\mathbb{R}$. Because A is measurable,

$$\begin{aligned} m^*(E \cap (A \cup B)) &= m^*((E \cap (A \cup B)) \cap A) + m^*((E \cap (A \cup B)) \cap A^c)) \\ &= m^*(E \cap A) + m^*(E \cap B \cap A^c). \end{aligned} \quad (3)$$

On the other hand,

$$\begin{aligned} m^*(E) &= m^*(E \cap A) + m^*(E \cap A^c) \\ &= m^*(E \cap A) + \left[m^*(E \cap A^c \cap B) + m^*(E \cap A^c \cap B^c)\right] \\ &= \left[m^*(E \cap A) + m^*(E \cap B \cap A^c)\right] + m^*(E \cap (A \cup B)^c) \\ &= m^*(E \cap (A \cup B)) + m^*(E \cap (A \cup B)^c). \end{aligned}$$

(At the next to last step we used $A^c \cap B^c = (A \cup B)^c$ and at the last step we used (3).) This shows that $A \cup B$ is measurable. Using Proposition 2 we also have A^c and B^c measurable, so both $A^c \cup B^c$ and $A \cap B = (A^c \cup B^c)^c$ are measurable. □

Proposition 10.4. *Suppose that $A_1, A_2, \ldots, A_N$ are pairwise disjoint measurable sets and suppose that E is a subset of $\mathbb{R}$. Then*

$$m^*\left(E \cap \bigcup_{k=1}^N A_k\right) = \sum_{k=1}^N m^*(E \cap A_k).$$

Proof: Let $B_n = \bigcup_{k=1}^n A_k$ for $1 \leq n \leq N$. By induction, each B_n is measurable. The desired equality is obtained by induction, using Proposition 3. In fact, if

$$m^*(E \cap B_n) = \sum_{k=1}^n m^*(E \cap A_k),$$

then

$$\begin{aligned} m^*(E \cap B_{n+1}) &= m^*\left(E \cap B_{n+1} \cap A_{n+1}^c\right) + m^*(E \cap B_{n+1} \cap A_{n+1}) \\ &= m^*(E \cap B_n) + m^*(E \cap A_{n+1}) = \sum_{k=1}^{n+1} m^*(E \cap A_k). \quad \square \end{aligned}$$

Proposition 10.5. *Suppose that $(A_n)_{n=1}^{\infty}$ is a sequence of pairwise disjoint measurable sets. Then the union $\bigcup_{n=1}^{\infty} A_n$ is measurable and*

$$m^*\left(\bigcup_{n=1}^{\infty} A_n\right) = \sum_{n=1}^{\infty} m^*(A_n). \tag{4}$$

Proof: Let $B_n = \bigcup_{k=1}^{n} A_k$ and let $B = \bigcup_{k=1}^{\infty} A_k$. For any $E \subset \mathbb{R}$, since B_N is measurable, we can use Proposition 10.4 to obtain

$$m^*(E) = m^*(E \cap B_N) + m^*(E \cap B_N^c) = \left[\sum_{n=1}^{N} m^*(E \cap A_n)\right] + m^*(E \cap B_N^c).$$

Now $B_N^c \supset B^c$, so we may take the limit as $N \to \infty$ to get

$$m^*(E) \geq \left[\sum_{n=1}^{\infty} m^*(E \cap A_n)\right] + m^*(E \cap B^c) \geq m^*(E \cap B) + m^*(E \cap B^c),$$

since $E \cap B = \bigcup_{n=1}^{\infty}(E \cap A_n)$. Therefore B is measurable. Moreover,

$$m^*(E) = m^*(E \cap B) + m^*(E \cap B^c) \leq \left[\sum_{n=1}^{\infty} m^*(E \cap A_n)\right] + m^*(E \cap B^c),$$

so we have equality:

$$m^*(E) = \left[\sum_{n=1}^{\infty} m^*(E \cap A_n)\right] + m^*(E \cap B^c). \tag{5}$$

Taking $E = B$ in (5), we obtain (4). □

Proposition 10.6. *Suppose that $(B_n)_{n=1}^{\infty}$ is any sequence of measurable sets. Then the intersection $\bigcap_{n=1}^{\infty} B_n$ and the union $\bigcup_{n=1}^{\infty} B_n$ are also measurable.*

Proof: Let $A_1 = B_1$, $A_2 = B_2 \cap B_1^c$, ..., $A_n = B_n \cap (B_1 \cup B_2 \cup \cdots \cup B_{n-1})^c$, and so on. Then it is easy to check that the sets A_n are pairwise disjoint and

$$\bigcup_{k=1}^{n} A_k = \bigcup_{k=1}^{n} B_k, \quad \text{all } n.$$

Propositions 10.2 and 10.3 imply that each A_n is measurable, and Proposition 10.5 gives the measurability of $\bigcup_{n=1}^{\infty} A_n = \bigcup_{n=1}^{\infty} B_n$.

For the intersection, we have

$$\bigcap_{n=1}^{\infty} B_n = \left(\bigcup_{n=1}^{\infty} B_n^c\right)^c.$$

Thus the intersection is also measurable. □

Exercises

1. Prove that $m^*(A) = 0$ implies that A is measurable.
2. Suppose that A is a subset of $\mathbb{R}$ with the property that for every $\varepsilon > 0$ there are measurable sets B and C such that
$$B \subset A \subset C, \qquad m(C \cap B^c) < \varepsilon.$$
Show that A is measurable
3. Select a sequence $(\varepsilon_n)_1^\infty$ of 0's and 1's at random in such a way that each ε_n has probability one-half of being 0. Let $x = \varepsilon_1/2 + \varepsilon_2/4 + \varepsilon_3/8 + \cdots$. Show that the probability that x lies in the subinterval $[a, b] \subset [0, 1]$ is $b - a$.

10D. Fundamental Properties of Measurable Sets

Suppose that A is a measurable set. Then its (Lebesgue) *measure* is defined to be its outer measure, denoted by m:

$$m(A) = m^*(A) \qquad \text{if } A \text{ is measurable.}$$

We have verified the first three of the following properties of measurable sets, and the remaining properties will be proved in this section.

I. *Complements, countable unions, and countable intersections of measurable sets are measurable.*
II. *Any interval is measurable, and its measure is its length.*
III. *Countable additivity*: If $(A_n)_{n=1}^\infty$ is a sequence of pairwise disjoint measurable sets, then
$$m\left(\bigcup_{n=1}^\infty A_n\right) = \sum_{n=1}^\infty m(A_n).$$
IV. *Continuity*: Suppose that $\{A_n\}_1^\infty$ and $\{B_n\}_1^\infty$ are sequences of measurable sets such that $A_1 \supset A_2 \supset A_3 \cdots$ and $B_1 \subset B_2 \subset B_3 \ldots$, and suppose that $m(A_1)$ is finite. Then
$$m\left(\bigcap_{n=1}^\infty A_n\right) = \lim_{n\to\infty} m(A_n), \quad m\left(\bigcup_{n=1}^\infty B_n\right) = \lim_{n\to\infty} m(B_n).$$
V. *Translation invariance*: If A is measurable and h is real, then the translate $A + h$ is measurable and $m(A + h) = m(A)$.
VI. *Open subsets and closed subsets* of $\mathbb{R}$ are measurable.
VII. *Approximation*: If A is measurable, then for any $\varepsilon > 0$ there exist a closed set B and an open set C such that
$$B \subset A \subset C, \qquad m(C \cap B^c) < \varepsilon.$$

In particular, $m(B) \geq m(A) - \varepsilon$ and $m(C) \leq m(A) + \varepsilon$. If $m(A)$ is finite, then B can be taken to be bounded.

Proof of continuity: With $(B_n)_{n=1}^\infty$ as in the statement, set

$$C_1 = B_1, \qquad C_{n+1} = B_{n+1} \cap B_n^c.$$

Then the C_n's are pairwise disjoint and $B_N = C_1 \cup C_2 \cup \cdots \cup C_N$. It follows that $\bigcup_{n=1}^\infty B_n = \bigcup_{n=1}^\infty C_n$ and so

$$m\left(\bigcup_{n=1}^\infty B_n\right) = m\left(\bigcup_{n=1}^\infty C_n\right) = \sum_{n=1}^\infty m(C_n) = \lim_{N\to\infty} \sum_{n=1}^N m(C_n) = \lim_{N\to\infty} m(B_N).$$

Next, suppose that $(A_n)_{n=1}^\infty$ is a sequence as in the statement. Let $A = \bigcap_{n=1}^\infty A_n$ and set $B_n = A_1 \cap A_n^c$. Then $B_1 \subset B_2 \subset \cdots$ and $\bigcup_{n=1}^\infty B_n = A_1 \cap A^c$. Moreover, $A_1 = A_n \cup B_n$ and $A_n \cap B_n = \emptyset$. Therefore

$$m(A_1) = m(A) + m(A_1 \cap A^c) = m(A) + \lim_{n\to\infty} m(B_n).$$

But also $m(A_1) = m(A_n) + m(B_n)$ and $m(A_n)$ has a limit since it is a bounded, nonincreasing sequence. Thus

$$m(A_1) = \lim_{n\to\infty} m(A_n) + \lim_{n\to\infty} m(B_n).$$

Since $m(A_1)$ is finite, it follows from these last two equations that $\lim m(A_n) = m(A)$. □

Proof of translation invariance: Suppose that A is measurable and E is a subset of $\mathbb{R}$. Note that

$$E \cap (A + h) = \big[(E - h) \cap A\big] + h, \qquad E \cap (A + h)^c = \big[(E - h) \cap A^c\big] + h.$$

It follows from the measurability of A and the translation invariance of m^* (see (5)) that $m^*(E) = m^*(E \cap (A + h)) + m^*(E \cap (A + h)^c)$, as desired. □

Proof that open and close sets are measurable: Since the complement of an open set is closed, we only need to consider open sets. Given A open and a point $x \in A$, choose an interval I_x having rational endpoints, such that $x \in I_x \subset A$. There are at most countably many *distinct* such intervals that arise, so they can be numbered as a sequence $(I_n)_{n=1}^\infty$, and A is the union of the sequence of measurable sets $(I_n)_{n=1}^\infty$. □

Proof of the approximation property: Suppose first that $A \subset J$, where J is a bounded closed interval. By definition, $m(A) = m^*(A)$, and so there are open intervals $(I_n)_{n=1}^\infty$ with $A \subset \bigcup_{n=1}^\infty I_n = C$ such that

$$m(A) \leq m(C) \leq \sum_{n=1}^\infty |I_n| < m(A) + \frac{\varepsilon}{2}.$$

Similarly, there is an open set $C' \supset J \cap A^c$ such that $m(C') < m(J \cap A^c) + \varepsilon/2$. Then $B = J \cap (C')^c$ is closed and $B \subset A$, while

$$m(J) = m(C') + m(B) < m(J \cap A^c) + m(B) + \varepsilon/2.$$

But also $m(J) = m(J \cap A^c) + m(A)$, so $m(A) < m(B) + \varepsilon/2$. Then

$$m(C \cap B^c) = m(C \cap A^c) + m(A \cap B^c) = m(C) - m(A) + m(A) - m(B) < \varepsilon.$$

Now we drop the assumption that A is bounded. Let $A_n = A \cap [n, n+1], n \in \mathbb{Z}$. By what has just been proved, there are closed sets B_n and open sets C_n such that

$$B_n \subset A_n \subset C_n, \qquad m(C_n \cap B_n^c) < \frac{\varepsilon}{2^{|n|+2}}.$$

Then $B = \bigcup_{n=-\infty}^\infty B_n$ is closed, $C = \bigcup_{n=-\infty}^\infty C_n$ is open, and these sets have the desired properties. If $m(A)$ is finite, then the continuity property implies that

$$\lim_{N \to \infty} m\left(B \setminus \bigcup_{n=-N}^N B_n\right) = 0.$$

Therefore we may replace the countable union B by the bounded set $\bigcup_{-N}^N B_n$ if N is taken to be sufficiently large. □

Exercises

1. Give an example of a sequence $(A_n)_{n=1}^\infty$ with $A_1 \supset A_2 \supset \cdots$ such that each $m(A_n)$ is infinite but $\bigcap_{n=1}^\infty A_n = \emptyset$.
2. Given a subset $E \subset \mathbb{R}$, let $E_n = E \cap [-n, +n]$ for $n \in \mathbb{Z}$. Show that $m^*(E) = \lim_{n \to \infty} m^*(E_n)$.
3. Suppose that $(A_n)_1^\infty$ is a sequence of subsets of $\mathbb{R}$. Define

   $$\limsup_{n \to \infty} A_n = \{x : x \text{ belongs to } A_n \text{ for infinitely many values of } n\}.$$

   $$\liminf_{n \to \infty} A_n = \{x : x \text{ belongs to } A_n \text{ for all but finitely many values of } n\}.$$

   (a) Give an example to show that these sets may be different.
   (b) Show that if each A_n is measurable, then so are $\liminf A_n$ and $\limsup A_n$. In fact, show that

   $$\liminf_{n \to \infty} A_n = \bigcup_{n=1}^\infty \left(\bigcap_{m=n}^\infty A_m\right),$$

   and find a similar expression for $\limsup A_n$.

4. The *symmetric difference* of sets A, B is defined to be the set

$$A \triangle B = (A \cap B^c) \cup (B \cap A^c)$$

of points belonging to one of the two sets but not both. Show that

$$d(A, B) = m(A \triangle B)$$

defines a *semi-metric* on the family $\mathcal{M}$ of all measurable subsets of $\mathbb{R}$ that have finite measure. (A semi-metric on a set X is a function d defined on pairs of elements of X such that $d(x, y) \geq 0$, $d(x, y) = d(y, x)$, and $d(x, z) \leq d(x, y) + d(y, z)$.)

5. Show that $(\mathcal{M}, d)$ is *complete*: If $(A_n)_1^\infty$ is a Cauchy sequence in $\mathcal{M}$, then there is a set $A \in \mathcal{M}$ such that $\lim_{n \to \infty} d(A_n, A) = 0$.

10E*. A Nonmeasurable Set

Is every subset of $\mathbb{R}$ measurable? Here is the standard counterexample. Define an equivalence relation for real numbers by setting

$$x \approx y \qquad \text{if } x - y \text{ is rational.}$$

This allows us to partition $\mathbb{R}$ into disjoint nonempty sets E_α, with

$$x, y \in E_\alpha \qquad \text{for some } \alpha \text{ if and only if } x \approx y.$$

For any real x, let $[x]$ denote the largest integer $\leq x$. Then

$$x \approx x - [x] \qquad \text{and} \qquad x - [x] \in [0, 1).$$

We "construct" a set A, using the Axiom of Choice, by choosing *one element from each set* $E_\alpha \cap [0, 1)$. Let $(x_n)_{n=1}^\infty$ be an enumeration of the rationals in the interval $(-1, 1)$ and set

$$A_n = A + x_n.$$

The sets $(A_n)_{n=1}^\infty$ are pairwise disjoint. In fact, if there were a point $x \in A_m \cap A_n$, then for some points x', x'' in A, $x = x' + x_m = x'' + x_n$, so $x' - x'' = x_m - x_n \in \mathbb{Q}$, so $x' \approx x''$, so $x_m = x_n$, and so $n = m$.
There are inclusions

$$(0, 1) \subset \bigcup_{n=1}^\infty A_n \subset (-1, 2).$$

In fact, by construction, for each $x \in (0, 1)$, there is a unique $x' \in A \subset [0, 1)$ such that $x - x' \in \mathbb{Q}$. Then $x - x' = x_n$ for some n, and so $x \in A_n$. The other inclusion is obvious, since $A \subset [0, 1)$ and $|x_n| < 1$, all n.

The set A is not measurable. In fact, suppose it were. Then the inclusions and the fact that the A_n are pairwise disjoint imply

$$1 = m((0,1)) \leq m\left(\bigcup_{n=1}^{\infty} A_n\right) = \sum_{n=1}^{\infty} m(A_n) \leq 3 = m([-1,2]).$$

But $m(A_n) = m(A + x_n) = m(A)$ for all n, so the preceding inequality is

$$1 \leq m(A) + m(A) + m(A) + \cdots \leq 3.$$

Neither $m(A) = 0$ nor $m(A) > 0$ is possible here, so A cannot be measurable.

11
Lebesgue Integration on the Line

We motivated the concept of measure in connection with integration. In this chapter we take advantage of the properties of measure to define and establish properties of the corresponding integration concept.

11A. Measurable Functions

If f is a function from $\mathbb{R}$ to $\mathbb{R}$ and A is a subset of $\mathbb{R}$, recall that the inverse image $f^{-1}(A)$ is the set defined by

$$f^{-1}(A) = \{x \in \mathbb{R} : f(x) \in A\}.$$

This mapping from sets to sets preserves complements and countable (indeed arbitrary) unions and intersections:

$$f^{-1}(A^c) = (f^{-1}(A))^c,$$

$$f^{-1}\left(\bigcup_{n=1}^{\infty} A_n\right) = \bigcup_{n=1}^{\infty}(f^{-1}A_n),$$

$$f^{-1}\left(\bigcap_{n=1}^{\infty} A_n\right) = \bigcap_{n=1}^{\infty}(f^{-1}A_n).$$

To motivate what follows, suppose that f is a bounded continuous function defined on the interval $I = [0, 1)$. Consider the problem of trying to integrate f; in fact, suppose that we want to compute the integral to a prescribed accuracy, say within $1/1000$. The standard method says: Cut the interval I into disjoint subintervals of length $\leq \delta$ and compute a corresponding Riemann sum or Darboux upper and lower sums. Then, for δ sufficiently small, the accuracy will be within $1/1000$ – *but how small must δ be?*

11A. Measurable Functions

Here is another method: Divide the *range* of f into disjoint intervals $I_j = [a_j, a_{j+1})$ of length $< 1/1000$ and let $A_j = f^{-1}(I_j)$. Then the function g that takes the value a_j on the set A_j comes within $1/1000$ of f on the whole interval I, so its integral should be within $1/1000$ that of f. In many cases each A_j is a union of finitely many disjoint intervals, so that

$$\int_0^1 g(x)\,dx = \sum_j a_j\, m(A_j). \tag{1}$$

We shall see that in all cases the continuity of f implies that the sets A_j are measurable and therefore the sum in (1) makes sense. Moreover, even if we do not assume continuity, this sum should give a good approximation to the integral – whatever we mean by the integral – so long as the sets A_j are measurable.

Recall that f is continuous if and only if $f^{-1}(A)$ is open whenever A is open. This fact, and the preceding remarks about integration, lead us to formulate the following definition of a "reasonable" function.

Definition. A real-valued function f defined on $\mathbb{R}$ is *measurable* if, for each open subset A of $\mathbb{R}$, the inverse image $f^{-1}(A)$ is measurable.

Note that if f is continuous, then it is measurable, since open sets are measurable. There are a number of *equivalent* formulations of this condition:

(i) $f^{-1}(A)$ is measurable whenever A is open.
(ii) $f^{-1}(A)$ is measurable whenever A is closed.
(iii) $f^{-1}(A)$ is measurable whenever A is an interval.
(iv) $f^{-1}(A)$ is measurable whenever A is an open interval.
(v) $f^{-1}(A)$ is measurable whenever A is an interval of the form $(a, +\infty)$.

In fact, (i) and (ii) are equivalent by taking complements, while (i) and (ii) imply (iii) because open intervals are open sets, closed intervals are closed sets, and a half-open interval is the intersection of an open interval and a closed interval. Obviously (iii) implies (iv) and (iv) implies (v). To complete the circle we show that (v) implies (i). Note that

$$(-\infty, b) = \bigcup_{n=1}^{\infty} \left(b - \tfrac{1}{n}, +\infty\right)^c, \qquad (a, b) = (-\infty, b) \cap (a, +\infty),$$

so (v) implies (iv). Any open set in $\mathbb{R}$ is a countable union of open intervals (see Exercise 6 of Section 10B), so (iv) implies (i) and the circle is complete.

Definition. If f and g are real-valued functions on $\mathbb{R}$, then the functions $f \wedge g$ and $f \vee g$ are defined by

$$(f \wedge g)(x) = \min\{f(x), g(x)\}, \qquad (f \vee g)(x) = \max\{f(x), g(x)\}.$$

Proposition 11.1. *Suppose that f and g are measurable functions, and suppose that c is real. Then the functions cf, $f+g$, fg, $|f|$, $f \wedge g$, and $f \vee g$ are measurable.*

Proof: We leave cf and fg as an exercise and check condition (v) for the remaining functions. It is convenient here and later to have a shorthand notation:

$$\{f > a\} = \{x \in \mathbb{R} : f(x) > a\} = f^{-1}(a, +\infty).$$

Then

$$\{|f| > a\} = \{f > a\} \cup \{f < -a\},$$
$$\{f \wedge g > a\} = \{f > a\} \cap \{g > a\},$$
$$\{f \vee g > a\} = \{f > a\} \cup \{g > a\},$$

so $|f|$, $f \wedge g$, and $f \vee g$ are measurable. As for the sum $f + g$, note that

$$f(x) > r, \quad g(x) > a - r \quad \text{implies } (f+g)(x) > a.$$

Conversely, if $(f+g)(x) > a$, then there is a *rational* r such that $f(x) > r > a - g(x)$. Therefore

$$\{f + g > a\} = \bigcup_{r \in \mathbb{Q}} (\{f > r\} \cap \{g > a - r\}),$$

which is a countable union of measurable sets. $\square$

It will be convenient to extend our notion of a measurable function somewhat, by allowing our functions to take the values $\pm\infty$ as well as values in $\mathbb{R}$. We may take (v) above as the definition, but with the interval A taken to include ∞. The preceding results do not change, although we must assume that the functions are such that the sum and product are *defined*; this is not automatic, because $\infty + (-\infty)$ and $0 \cdot (\pm\infty)$ are not defined.

Proposition 11.2. *Suppose that $(f_n)_{n=1}^{\infty}$ is a sequence of measurable functions and suppose that, for each real x, $\lim_{n \to \infty} f_n(x) = f(x)$. Then f is measurable.*

Proof: Suppose that $f(x) > a$. Then for some $m \in \mathbb{N}$, $f(x)$ is greater than $a + (1/m)$. Therefore there is N such that $f_n(x) > a + (1/m)$ for all $n \geq N$. This

shows
$$\{f > a\} \subset \bigcup_{m=1}^{\infty} \bigcup_{N=1}^{\infty} \bigcap_{n=N}^{\infty} \left\{f_n > a + \frac{1}{m}\right\}.$$

Conversely, if x belongs to the set on the right, then for some m and some N we have $f_n(x) > a + (1/m)$ for all $n \geq N$, so the limit $f(x)$ is $\geq a + (1/m) > a$. Therefore the two sets above are equal and thus $\{f > a\}$ is measurable. □

Proposition 11.3. *Suppose that $(f_n)_{n=1}^{\infty}$ is a sequence of measurable functions. Let*
$$g(x) = \inf_n \{f_n(x)\}, \quad h(x) = \sup_n \{f_n(x)\}.$$
Then g and h are measurable.

(It is in results like this that it is convenient to allow the values $\pm\infty$, so that g and h are always defined at every point.)

Proof: For any $a \in \mathbb{R}$,
$$\{g < a\} = \bigcup_{n=1}^{\infty} \{f_n < a\}, \quad \{h > a\} = \bigcup_{n=1}^{\infty} \{f_n > a\}. \quad \Box$$

Exercises

1. Prove that the indicator function 1_A of a subset A of $\mathbb{R}$ is measurable if and only if the set A is measurable.
2. Suppose that f is a function from $\mathbb{R}$ to the extended reals, $\mathbb{R} \cup \{-\infty, +\infty\}$. Reinterpret criterion (v) for such a function. Show that if f is measurable according to the new criterion (v), then the sets $\{f = -\infty\}$ and $\{f = +\infty\}$ are measurable.
3. Suppose that f and g are real-valued functions on $\mathbb{R}$ such that f is continuous and g is measurable. Prove that the composition $f(g)$ is measurable.
4. Suppose that f and g are measurable real-valued functions.
   (a) Suppose that A is an open subset of $\mathbb{R}^2$. Prove that $\{x \in \mathbb{R} : (f(x), g(x)) \in A\}$ is a measurable set.
   (b) Suppose that H is a continuous real-valued function on $\mathbb{R}^2$. Prove that the function h defined by $h(x) = H(f(x), g(x))$ is measurable.
   (c) Deduce from (b) that the sum $f + g$ and the product fg are measurable.
5. As hinted at in the discussion about integration, $f^{-1}(I)$ may not consist of finitely many disjoint intervals, even when f is a continuous function defined on a closed bounded interval and I is an interval. Give an example.
6. Prove *Egorov's Theorem*: Suppose that $\{f_n\}$ is a sequence of measurable functions that converge to f at each point of a set A that has finite measure. Prove that for any positive

ε there is a set $B \subset A$ such that $m(B) < \varepsilon$ and f_n converges to f *uniformly* on the difference set $A \setminus B = A \cap B^c$.

11B*. Two Examples

It might seem that a more natural requirement for the measurability of a function f would be that $f^{-1}(A)$ be measurable for every *measurable* A. This turns out to be too restrictive. Let us define $g : [0, 1] \to [0, 1]$ as follows. Any number x in the interval $[0, 1]$ has a binary expansion

$$x = \sum_{n=1}^{\infty} \frac{\varepsilon_n}{2^n}, \quad \text{each } \varepsilon_n = 0 \text{ or } 1.$$

This expansion is unique if we take the expansion that terminates whenever possible, that is, when $x = p/2^n$ for some integers p, n. Then set

$$g(x) = \sum_{n=1}^{\infty} \frac{2\varepsilon_n}{3^n}.$$

This function is *strictly increasing, continuous except at the points $p/2^n$ for $0 < p \le 2^n$, and maps $[0, 1]$ to the Cantor set C*. We extend g by taking $g(x) = 0$ if $x \notin [0, 1]$. Suppose now that B is a subset of $(0, 1]$ that is not measurable and let $A = g(B)$. Then $A \subset C$, so $m^*(A) = 0$ and thus A is measurable. Now g is strictly increasing. Therefore g is 1–1 on $(0, 1]$ and $g^{-1}(A) = B$. But g is not an evil function. In fact, it is a pointwise limit of the step functions

$$g_n(x) = \sum_{k=1}^{n} \frac{2\varepsilon_n}{3^n}.$$

Thus, to require that inverse images of measurable sets be measurable would be to rule out even some limits of step functions.

The function g above is closely related to another interesting example. Define f by taking the ternary expansion of $x \in [0, 1]$:

$$x = \sum_{n=1}^{\infty} \frac{\varepsilon_n}{3^n}, \quad \varepsilon_n = 0, 1, \text{ or } 2.$$

We opt for the terminating expansion if and only if the last (nonzero) digit of the terminating expansion is 2. If x belongs to the Cantor set C, that is, if no $\varepsilon_n = 1$, we define

$$f(x) = \frac{1}{2} \sum_{n=1}^{\infty} \frac{\varepsilon_n}{2^n}.$$

On the complement, which is the union of the successive middle third intervals, we define f on an interval to be its value at the endpoints (f takes the same value at each endpoint). Thus $f = 1/2$ on the interval $(\frac{1}{3}, \frac{2}{3})$, $f = 3/4$ on $(\frac{7}{9}, \frac{8}{9})$, and so on. This function f has the properties

f maps [0,1] onto [0,1];
f is nondecreasing;
$f'(x)$ exists and $= 0$ except on a set having measure 0.

In fact, clearly $f'(x) = 0$ for $X \in [0, 1] \cap C^c$. Except at countably many points of $[0, 1]$, $f(g(x)) = x$.

Exercises

1. Prove that the function g in Section 11B is continuous except at the points $p/2^n$.
2. Show that the composition of two measurable functions may not be measurable.

11C. Integration: Simple Functions

A starting point for the theory of the (definite) Riemann integral is the integral of a *step function*, that is, a function that is constant on intervals, using the geometric notion of the length of an interval. We have extended the latter notion to a much larger collection of sets, so we have a much larger collection of basic functions at our disposal.

Definition. The *indicator function* of a subset A of $\mathbb{R}$ is the function

$$1_A(x) = 1 \quad \text{if } x \in A, \qquad = 0 \quad \text{if } x \notin A.$$

Definition. An *integrable simple function* (ISF) is a function of the form

$$f = a_1 1_{A_1} + \cdots + a_n 1_{A_n}, \qquad (2)$$

where the a_j's are real numbers and the A_j's are disjoint measurable sets having finite measure.

Definition. The *integral* of an ISF of the form (1) is

$$\int f = \int_{\mathbb{R}} f = \int_{\mathbb{R}} f(x)\,dx = \sum_{k=1}^{n} a_k\, m(A_k). \qquad (3)$$

Strictly speaking, we should first verify that the sum in (3) depends only on the *function f* and not on the particular way of expressing it in the form (2). Suppose in fact that

$$\sum_{j=1}^{m} a_j 1_{A_j} \equiv \sum_{k=1}^{n} b_k 1_{B_k}, \qquad (4)$$

where the A_j's, and B_k's are measurable sets with finite measure, while the sets $\{A_j\}$ are pairwise disjoint, as are the sets $\{B_k\}$. The equality (4) then implies that $a_j = b_k$ if $A_j \cap B_k$ is not empty. Therefore

$$\sum_{j=1}^{m} a_j m(A_j) = \sum_{j=1}^{m} \left(\sum_{k=1}^{n} a_j m(A_j \cap B_k) \right) = \sum_{j=1}^{m} \left(\sum_{k=1}^{m} b_k m(A_j \cap B_k) \right)$$

$$= \sum_{k=1}^{n} b_k \left(\sum_{j=1}^{m} m(A_j \cap B_k) \right) = \sum_{k=1}^{n} b_k m(B_k).$$

Here is a collection of basic properties of integrable simple functions. The verifications, which are easy, are left as an exercise.

I. *If f and g are ISFs and c is real, then cf and $f + g$ are ISFs and*

$$\int cf = c \int f, \qquad \int (f+g) = \int f + \int g.$$

Moreover, $|f|$ is an ISF and

$$\left| \int f \right| \leq \int |f|.$$

II. *If f and g are integrable simple functions and $f \leq g$, then $\int f \leq \int g$.*

III. *If f is an ISF and f_a is its translate by $a \in \mathbb{R}$, $f_a(x) = f(x - a)$, then f_a is an ISF and*

$$\int f_a = \int f.$$

Example. Let f be the function mentioned in the introduction to Chapter 10, extended to the line: $f(x) = 1$ if x is rational, $f(x) = 0$ if x is irrational. Then $f = 1_\mathbb{Q}$ is an ISF with integral $\int f = m(\mathbb{Q}) = 0$, since $\mathbb{Q}$ is countable.

Exercise

1. Verify the properties I, II, and III of integrable simple functions.

11D. Integration: Measurable Functions

At this point we can introduce a notion of the integral for any *nonnegative* measurable function.

Definition. If $f : \mathbb{R} \to \mathbb{R}$ is a nonnegative measurable function, then $\int f$ is the supremum

$$\sup\left\{\int g : g \text{ is an ISF such that } 0 \le g \le f\right\}. \tag{5}$$

(In particular, it is possible that $\int f = +\infty$.)

Notice that there is no conflict in notation, because if f is itself an ISF, then, for any g as in (5), $\int g \le \int f$; on the other hand, in this case we can take $g = f$, so the supremum is $\int f$ as defined earlier.

Once again we begin with basic properties.

If f and g are nonnegative measurable functions and $a > 0$, then

$$\int af = a\int f, \qquad \int (f+g) = \int f + \int g.$$

Moreover, if $f \le g$, then $\int f \le \int g$.

The only one of these statements that is not fairly obvious is additivity. The proof of additivity requires a result on approximation that is of interest in itself. (The idea here is the one we already used at the beginning of Section 11A in discussing how to approximate a Riemann integral.)

Lemma 11.4. *Suppose that f is a bounded measurable function and A is a set with finite measure. For any $\varepsilon > 0$, there are integrable simple functions f_1 and f_2 such that $f_1 \le f \le f_2$ on A and $f_2 - f_1 \le \varepsilon$ at every point.*

Proof: Choose M so that $|f(x)| \le M$ for all x's and decompose the interval $[-M, +M]$ into disjoint subintervals $I_1, I_2, \ldots, I_n$ of length $\le \varepsilon$. Let $A_k = A \cap f^{-1}(I_k)$ and let

$$f_1 = \sum_{k=1}^{n} a_k 1_{A_k}, \qquad f_2 = \sum_{k=1}^{n} b_k 1_{A_k},$$

where a_k and b_k are the left and right endpoints of the interval I_k. □

Proof of additivity: Suppose that f and g are nonnegative measurable functions, and suppose that f_1, g_1 are ISFs such that $0 \le f_1 \le f$ and $0 \le g_1 \le g$. Then $f_1 + g_1 \le f + g$, so

$$\int f_1 + \int g_1 = \int (f_1 + g_1) \le \int (f+g).$$

(We used additivity of the integral for ISFs.) Take the supremum over all such ISFs to deduce that

$$\int f + \int g \leq \int (f+g).$$

To prove the reverse inequality, suppose that h is an ISF and $0 \leq h \leq f+g$. Let A be the set where h is positive; it has finite measure. Now h is bounded, so the functions $h \wedge f$ and $h \wedge g$ are bounded. We can use the lemma to choose ISFs f_1, g_1 such that

$$0 \leq f_1 \leq h \wedge f \leq f_1 + \varepsilon, \qquad 0 \leq g_1 \leq h \wedge g \leq g_1 + \varepsilon.$$

Now $h \leq f+g$, so $h \leq h \wedge f + h \wedge g \leq f_1 + g_1 + 2\varepsilon 1_A$ and so

$$\int h \leq \left(\int f_1 + \int g_1\right) + 2\varepsilon\, m(A) \leq \left(\int f + \int g\right) + 2\varepsilon\, m(A).$$

Taking first the infimum over $\varepsilon > 0$ and then the supremum over $h \leq f+g$, we obtain the remaining inequality

$$\int \int (f+g) \leq \int f + \int g. \qquad \square$$

The additivity property will be important as we extend the definition of integration to functions that change signs. For any real-valued or extended-real-valued function f on R, define nonnegative functions

$$f^+(x) = f(x) \text{ if } f(x) > 0, \qquad f^+(x) = 0 \text{ if } f(x) \leq 0;$$
$$f^-(x) = -f(x) \text{ if } f(x) < 0, \qquad f^-(x) = 0 \text{ if } f(x) \geq 0.$$

These functions are measurable if f is, and

$$f = f^+ - f^-, \qquad |f| = f^+ + f^-.$$

Definition. A measurable function f is *integrable* if $\int |f| < +\infty$. Note that functions $f^\pm$ satisfy $0 \leq f^\pm \leq |f|$, so f is integrable if and only if $\int f^+ < +\infty$ and $\int f^- < \infty$.

Definition. If f is integrable, its *integral* $\int f$ is defined to be $\int f^+ - \int f^-$. (This is consistent with the earlier definition, since $f \geq 0$ implies $f^+ = f$ and $f^- = 0$.)

Basic properties of integrability and the integral

I. *Suppose that f and g are integrable and a is real. Then af and $f+g$ are integrable and $\int af = a \int f$, $\int (f+g) = \int f + \int g$.*

II. If f and g are integrable and $f \leq g$, then $\int f \leq \int g$. Also, $|\int f| \leq \int |f|$.
III. If f is integrable and f_a is its translate by $a \in \mathbb{R}$, then f_a is integrable and

$$\int f_a = \int f.$$

These properties require some proof. We concentrate on additivity, which is not obvious, since in general $(f+g)^{\pm} \neq f^{\pm} + g^{\pm}$. First, note that $|f+g| \leq |f| + |g|$; so if f and g are integrable, so is $f+g$. Next, note that

$$(f+g)^+ - (f+g)^- = f+g = (f^+ - f^-) + (g^+ - g^-).$$

Therefore

$$(f+g)^+ + f^- + g^- = (f+g)^- + f^+ + g^+.$$

All the functions in the preceding identity are nonnegative, so

$$\int (f+g)^+ + \int f^+ + \int g^- = \int (f+g)^- + \int f^+ + \int g^+.$$

This identity implies

$$\int (f+g)^+ - \int (f+g)^- = \left[\int f^+ - \int f^- \right] + \left[\int g^+ - \int g^- \right],$$

so $\int (f+g) = \int f + \int g$.

Now $f \leq g$ implies $g - f \geq 0$, so $\int g - \int f = \int (g-f) \geq 0$. Since $-|f| \leq f \leq |f|$, we have also $|\int f| \leq \int |f|$. Property III follows from the corresponding property of ISFs.

So far we have defined integration only over the entire line. The restriction to subsets is easy.

Definitions. Suppose that A is a measurable set. A function f defined on A is measurable if $f^{-1}((a, +\infty])$ is measurable for every $a > 0$. Equivalently, f is measurable if the function $\tilde{f}$, obtained by extending f to vanish on the complement A^c, is measurable. The function f is said to be integrable if $\tilde{f}$ is integrable; if so, the integral $\int_A f$ is defined to be $\int \tilde{f}$.

If f is an integrable function on $\mathbb{R}$, the integral of f over A is the integral of the restriction of f to A, or, equivalently, $\int f 1_A$. (In the product $f 1_A$ we violate an earlier stricture and use the convention that $0 \cdot (\pm \infty) = 0$. Thus $f 1_A \equiv 0$ on A^c.)

Exercises

The functions in Exercises 1–9 are assumed to be nonnegative, measurable functions on $\mathbb{R}$. It may be be assumed that if f_n is a sequence of such functions with the properties

(i) $f_1 \leq f_2 \leq \cdots \leq f_n \ldots$;
(ii) $\lim_{n \to \infty} f_n(x) = f(x)$, all x;
(iii) each f_n vanishes outside a bounded closed interval $I_n = [a_n, b_n]$ and is continuous on that interval, where $\{I_n\}$ is an increasing sequence with union I,

then the integral of f is the limit of the Riemann integrals:

$$\int_I f = \lim_{n \to \infty} \int_{a_n}^{b_n} f_n(x) \, dx.$$

(This will follow from the results in Sections 11E and 12B.)

1. For what values of the exponent $a > 0$ is $\int f < \infty$, where $f(x) = x^{-a}$ if $0 < x < 1$ and $f(x) = 0$ otherwise?
2. For what values of the exponent $b > 0$ is $\int f < \infty$, where $f(x) = x^{-b}$ if $x > 1$ and $f(x) = 0$ otherwise?
3. Find a function f that vanishes outside the interval $[0, 1]$, such that $\int f$ is finite but $\int f^2$ is not finite.
4. Show that if f is bounded and $\int f < \infty$, then $\int f^2 < \infty$.
5. Find a bounded function g such that $\int g^2 < \infty$ but $\int g = \infty$.
6. Show that if g vanishes outside a bounded interval and $\int g^2 < \infty$, then $\int g < \infty$.
7. Suppose that $\int f < \infty$. For any $a > 0$, let $E_a = \{x : f(x) > a\}$. Prove that

$$m(E_a) \leq \frac{1}{a} \int f.$$

8. Find a function f such that $m(E_a) \leq 1/a$, all $a > 0$, where E_a is as in the preceding exercise but $\int f = +\infty$.
9. Suppose that f is continuous and $\int f < +\infty$. Is it always true that

$$\lim_{|x| \to \infty} f(x) = 0?$$

10. Suppose that $\int f_n \to 0$, and suppose that $\varepsilon > 0$ is given. Prove that

$$\lim_{n \to \infty} m(\{x : f_n(x) > \varepsilon\}) = 0.$$

11. Let $f(x) = \sin x / x$ for $x \neq 0$. Show that the limit

$$\lim_{N \to \infty} \int_{-N}^{N} f(x) \, dx$$

exists and is finite but f is not an integrable function.

11E. Convergence Theorems

The first theorem proved in this section is the single most important convergence theorem of integration theory. We begin with a very special case.

Lemma 11.5. *Suppose that g is a nonnegative ISF and suppose that $(f_n)_{n=1}^\infty$ is a sequence of measurable functions such that*

$$g \geq f_1 \geq f_2 \geq \cdots \geq f_n \geq \cdots \geq 0; \qquad \lim_{n \to \infty} f_n(x) = 0, \text{ all } x.$$

Then $\lim_{n \to \infty} \int f_n = 0$.

Proof: Since g is an ISF, it is bounded; say $0 \leq g(x) \leq M$, all x. Moreover, the set $A = \{g > 0\}$ has finite measure. Given $\varepsilon > 0$, let $A_n = \{f_n > \varepsilon\}$. Our assumptions imply

$$A \supset A_1 \supset A_2 \supset \cdots; \qquad \bigcap_{n=1}^\infty A_n = \emptyset.$$

Therefore (by continuity) $m(A_n) \to 0$. Choose N so that $m(A_n) < \varepsilon$ for $n \geq N$. Then $n \geq N$ implies

$$f_n = 0 \text{ on } A^c, \qquad f_n \leq \varepsilon \text{ on } A \cap A_n^c, \qquad f_n \leq M,$$

so

$$0 \leq \int f_n = \int_{A_n} f_n + \int_{A \cap A_n^c} f_n \leq M\, m(A_n) + \varepsilon\, m(A \cap A_n^c) < \varepsilon[M + m(A)]. \qquad \square$$

Theorem 11.6: Lebesgue's Dominated Convergence Theorem (DCT). *Suppose that $(f_n)_{n=1}^\infty$ is a sequence of measurable functions such that, for all real x, $\lim f_n(x) = f(x)$. Suppose also that the f_n's are dominated by an integrable function g, in the sense that, for every real x, $|f_n(x)| \leq g(x)$. Then*

$$\lim_{n \to \infty} \int f_n = \int f.$$

Proof: We consider three cases, of increasing generality.

Case 1. Suppose that $f = 0$ and $(f_n)_{n=1}^\infty$ is a nonincreasing sequence. Given $\varepsilon > 0$, choose an ISF g_1 with $0 \leq g_1 \leq g$ and $\int g \leq \int g_1 + \varepsilon$. Now

$$f_n = f_n \wedge g_1 + [f_n - f_n \wedge g_1] \leq f_n \wedge g_1 + (g - g_1).$$

The functions $f_n \wedge g_1$ satisfy the conditions of the lemma, since they are dominated by the ISF g_1. Therefore, for large n,

$$\int f_n \leq \int f_n \wedge g_1 + \int (g - g_1) \leq \varepsilon + \varepsilon.$$

Case 2. $f = 0$ and $f_n \geq 0$, all n. Let $g_n = \sup\{f_n, f_{n+1}, f_{n+2}, \ldots\}$. Then the g_n's decrease to 0 and Case 1 applies to the g_n, so

$$0 \leq \int f_n \leq \int g_n \to 0.$$

Case 3. The general case. The functions $g_n = |f_n - f|$ are nonnegative, measurable, converge to 0 at each point, and $g_n \leq |f_n| + |f| \leq 2g$, so Case 2 applies to the g_n and

$$0 \leq \left| \int f_n - \int f \right| \leq \int |f_n - f| = \int g_n \to 0.$$

This completes the proof. □

One easy and useful consequence of the DCT is the following.

Theorem 11.7: Monotone Convergence Theorem. *Suppose that $(f_n)_{n=1}^\infty$ is a nondecreasing sequence of nonnegative measurable functions: $0 \leq f_1 \leq f_2 \cdots$. Let $f(x) = \lim_{n \to \infty} f_n(x)$. Then $\lim_{n \to \infty} \int f_n = \int f$.*

Proof: Clearly $\lim_{n \to \infty} \int f_n \leq \int f$. If h is any ISF such that $0 \leq h \leq f$, then the Dominated Convergence Theorem implies that

$$\int h = \lim \int h \wedge f_n \leq \lim \int f_n.$$

Taking the supremum over such h, we obtain $\int f \leq \lim_{n \to \infty} \int f_n$. Thus $\int f = \lim_{n \to \infty} \int f_n$. □

The next result is an easy consequence of the Monotone Convergence Theorem. The proof is left as an exercise.

Theorem 11.8. Fatou's Lemma. *If $(f_n)_{n=1}^\infty$ is a sequence of nonnegative measurable functions, then*

$$\int \liminf_{n \to \infty} f_n \leq \liminf_{n \to \infty} \int f_n.$$

11E. Convergence Theorems

The three preceding results – Dominated, Monotone, and Fatou – are the fundamental convergence theorems of integration theory.

Exercises

1. Show that the following limits exist, and compute them (in (c), assume $a > 0$):

    (a) $\displaystyle\lim_{n\to\infty} \int_0^1 \frac{n^2 x^2}{e^{nx}}\, dx.$

    (b) $\displaystyle\lim_{n\to\infty} \int_0^1 \frac{1+nx}{(1+x)^n}\, dx.$

    (c) $\displaystyle\lim_{n\to\infty} \int_0^\infty \left(1+\frac{x}{n}\right)^n e^{-ax}\, dx.$

2. Suppose that f is integrable, and let $E(n)$ be the set $\{|f| > n\}$. Improve on Exercise 7 of Section 11D by showing that
$$\lim_{n\to\infty} n\, m(E_n) = 0.$$

3. Suppose that f is integrable and define $F(x) = \int_{(-\infty, x)} f$.
    (a) Prove that F is continuous.
    (b) Is F necessarily uniformly continuous?

4. Suppose that $0 < p < q < r$. Suppose that f is a nonnegative measurable function such that $\int f^p$ and $\int f^r$ are finite. Prove that $\int f^q$ is finite.

5. Prove Fatou's Lemma.

6. Show by an example that strict inequality is possible in the conclusion of Fatou's Lemma.

7. Suppose that f_n is a sequence of real-valued continuous functions on the interval $[0, 1]$ such that $|f_n(x)| \leq 1$, all $x \in [0, 1]$, and $\lim_{n\to\infty} f_n(x) = 0$, all $x \in [0, 1]$. By extending the f_n to vanish outside the interval, one can use the DCT to prove that $\int f_n \to 0$. Give a direct proof of this fact using some ideas from the proof of the DCT but not the DCT itself. (Although this is purely a theorem about the Riemann integral, it is difficult to find a proof that does not borrow ideas from measure theory.)

8. Suppose that $(f_n)_{n=1}^\infty$ is a sequence of measurable functions such that
$$0 \leq f_1 \leq f_2 \leq f_3 \leq \cdots$$
and let $f(x) = \lim f_n(x)$. Thus f may take the value $+\infty$ at some or all points. Prove that if $\lim \int f_n$ is finite, then f is finite except on a set of measure 0.

9. Find a sequence $(f_n)_{n=1}^\infty$ of nonnegative measurable functions such that
$$\lim_{n\to\infty} \int f_n = 0, \quad \limsup_{n\to\infty} f_n(x) = +\infty, \quad \text{all } x.$$

12

Integration and Function Spaces

In this chapter we establish the connection between integrability in the sense of Riemann and integrability in the sense of Lebesgue, introduce two important spaces of integrable functions, and establish a very general connection between integration and differentiation. Each of these developments depends on the notion of "almost everywhere."

12A. Null Sets and the Notion of "Almost Everywhere"

From the point of view of integration, sets of measure 0 are negligible. From the point of view of spaces of integrable functions, they are a (very) minor nuisance.

Definition. A *null set* is a measurable set with measure 0.

As an exercise from the definition of measurability, note that a set of outer measure 0 is measurable; therefore, *a subset of a null set is a null set.*

Definition. Two functions f and g are said to be equal *almost everywhere* if they coincide except on a null set. This is usually abbreviated as $f = g$ a.e. Another way to express this is that $f(x) = g(x)$ for almost every x, also abbreviated as a.e. x.

Proposition 12.1. *Suppose that f is a measurable function and that g is a function such that $g = f$ a.e. Then g is measurable.*

If f is nonnegative and measurable, then $\int f = 0$ if and only if $f = 0$ a.e. Also, if $\int f$ is finite, then f is finite a.e. In particular, any integrable function is finite a.e.

Proof: If $f = g$ a.e., then it is easy to see that, for any real a, the sets $\{f > a\}$ and $\{g > a\}$ differ by a null set, so the second set is measurable if the first is.

Suppose that $f \geq 0$. Let $A_n = \{f > \frac{1}{n}\}$. Then

$$\{f > 0\} = \bigcup_{n=1}^{\infty} A_n; \qquad \frac{1}{n} \cdot m(A_n) \leq \int_{A_n} f \leq \int f.$$

Therefore $\int f = 0$ implies $m(A_n) = 0$, all n, so $m(\{f > 0\}) = 0$.

Finally, let $B_n = \{f > n\}$, so that $\{f = +\infty\} = \bigcap_{n=1}^{\infty} B_n$. Then

$$n \cdot m(B_n) \leq \int_{B_n} f \leq \int f.$$

If $\int f$ is finite, then this implies $m(B_n) \to 0$, so $\{f = +\infty\}$ has measure 0. $\square$

Remark. Various hypotheses made earlier can be weakened. For example, in the Dominated Convergence Theorem, one only needs $\lim f_n(x) = f(x)$ a.e. and $|f_n(x)| \leq g(x)$ a.e.

12B*. Riemann Integration and Lebesgue Integration

As before, by a *step function* on an interval [a,b] we mean a function that is constant on subintervals. Setting such a function equal to 0 outside the interval, we obtain an ISF whose (Lebesgue) integral is clearly the same as the (Riemann) integral of the step function. Here we show that this equality extends to *all* Riemann integrable functions.

It is not difficult to deduce one more criterion for Riemann integrability from Proposition 8.16: A bounded function $f : [a, b] \to \mathbb{R}$ is Riemann integrable if and only if there are sequences of step functions $(g_n)_{n=1}^{\infty}$ and $(h_n)_{n=1}^{\infty}$ such that

$$g_n \leq f \leq h_n, \qquad \lim_{n \to \infty} \int (h_n - g_n) = 0.$$

Then the common value $\lim_{n \to \infty} \int g_n = \lim_{n \to \infty} \int h_n$ is the Riemann integral $\int_a^b f$.

Theorem 12.2. *If the real-valued function f is Riemann integrable on $I = [a, b]$, then it is Lebesgue integrable on I and the Lebesgue integral $\int_I f$ equals the Riemann integral $\int_a^b f$.*

Proof: The main trick is to prove that f is measurable. Choose sequences $(g_n)_{n=1}^{\infty}$, $(h_n)_{n=1}^{\infty}$ as above. We may replace these functions by

$$g_1 \vee g_2 \vee \cdots \vee g_n, \qquad h_1 \wedge h_2 \wedge \cdots \wedge h_n$$

and assume that the sequences are monotone:

$$g_1 \leq g_2 \leq \cdots \leq g_n \leq f \leq h_n \leq \cdots \leq h_2 \leq h_1.$$

Then the limits $g(x) = \lim_{n\to\infty} g_n(x)$ and $h(x) = \lim_{n\to\infty} h_n(x)$ exist, for each $x \in [a, b]$, and $g \leq f \leq h$. By dominated convergence,

$$\int_I g = \lim_{n\to\infty} \int_I g_n = \lim_{n\to\infty} \int_I h_n = \int_I h,$$

so $\int_I (h - g) = 0$. It follows from this and the inequality $g \leq h$ that $g = h$ almost everywhere (see Proposition 12.1). Therefore f is equal a.e. to the measurable functions g and h and is itself measurable. The argument shows that the Riemann and Lebesgue integrals of f are each equal to $\lim_{n\to\infty} \int_I g_n$. □

An example. The relationship between the Riemann and Lebesgue integrals does not hold for *improper* integrals. For example, the improper Riemann integral

$$\int_0^\infty \frac{\sin x}{x} dx = \lim_{n\to\infty} \int_0^n \frac{\sin x}{x} dx$$

exists but

$$\lim_{n\to\infty} \int_0^n \frac{|\sin x|}{x} dx = \infty,$$

so $\sin x / x$ is not Lebesgue integrable on $(0, \infty)$.

We know now that a Riemann integrable function is (bounded and) measurable.

Theorem 12.3. *A bounded, measurable real-valued function f on $[a, b]$ is Riemann integrable if and only if its points of discontinuity are a set of measure 0.*

Proof: The idea is to relate the Riemann integrability to the amount of oscillation of f over small distances. Set $A = [a, b]$, and for each x in A set

$$\omega_n(x) = \sup\left\{|f(y') - f(y)| : y, y' \in A, |y - x| < \frac{1}{n}, |y' - x| < \frac{1}{n}\right\}.$$

Then $\omega_1 \geq \omega_2 \geq \ldots$ and the ω_n's are measurable (why? – see Exercise 2). The function f is continuous at x if and only if $\lim_{n\to\infty} \omega_n = 0$. What we need to show is that Riemann integrability of f is equivalent to

$$\lim_{n\to\infty} m(\{\omega_n > \varepsilon\}) = 0, \quad \text{for all } \varepsilon > 0.$$

Suppose that the preceding limit is 0. Given $\varepsilon > 0$, choose n so large that the measure of the set $\{\omega_n > \varepsilon\}$ is less than ε. Divide A into finitely many disjoint subintervals A_j, each of length $\leq 1/n$. Define the usual largest step function g and smallest step function h such that $g \leq f \leq h$, while g and h are constant on the intervals A_j. If $h - g > \varepsilon$ on A_j, then $\omega_n > \varepsilon$ on A_j, so the total length of such intervals A_j is less than ε. If $|f| \le M$, then

$$\int_A (h-g) = \int_{(h-g)<\varepsilon} (h-g) + \int_{(h-g)>\varepsilon} (h-g) \le \varepsilon \cdot \big[m(A) + M\big].$$

Therefore f is Riemann integrable.

Conversely, suppose that f is Riemann integrable. Given $\varepsilon > 0$ and $\delta > 0$, choose step functions g, h such that

$$g \le f \le h, \qquad \int_A (h-g) \le \varepsilon\delta.$$

Assume that g and h are constant on each of the disjoint subintervals $A_1, \ldots, A_k$ with union A. Define the distance from a point x to a set B to be $d(x, B) = \inf\{|x - y| : y \in B\}$. Given a positive integer n set

$$A'_j = \left\{x \in A_j : d(x, A_j^c) < \frac{1}{n}\right\},$$
$$A''_j = \left\{x \in A_j : d(x, A_j^c) \ge \frac{1}{n} \text{ and } \omega_n(x) > \varepsilon\right\}.$$

It follows from these definitions that

$$\{\omega_n > \varepsilon\} \subset \bigcup_{j=1}^{k} (A'_j \cup A''_j).$$

If A''_j is not empty, then $h - g > \varepsilon$ on A_j, so

$$m(A''_j) \le \frac{1}{\varepsilon} \int_{A''_j} (h-g).$$

Moreover, the definitions imply that $m(A'_j) \le 2/n$. Therefore

$$m(\{\omega_n > \varepsilon\}) \le \sum_{j=1}^{k} m(A'_j) + \sum_{j=1}^{k} m(A''_j) \le \frac{2k}{n} + \frac{1}{\varepsilon} \int_A (h-g) \le \frac{2k}{n} + \delta.$$

Since n and δ are arbitrary, the proof is complete. □

Exercises

1. Prove that a bounded real-valued function f on $[a, b]$ is Riemann integrable in the sense of Section 8C if and only if there are sequences $\{g_n\}, \{h_n\}$ as in the proof of Theorem 12.2.
2. Let f and ω_n be as in Theorem 12.3. Show that $\omega_n^{-1}((a, \infty))$ is an open subset of the interval A, that is, if $\omega_n(x) > a$, then there is $\delta > 0$ such that $x' \in A$ and $|x' - x| < \delta$ implies $\omega_n(x') > a$.

3. Prove directly from the definitions that the indicator function of the Cantor set is Riemann integrable.
4. Suppose that A is a measurable subset of the interval $[a, b]$. Prove that the indicator function 1_A is Riemann integrable if and only if its boundary ∂A (see Exercise 10 of Section 6B) has measure 0.

12C. The Space L^1

The *function space* $L^1 = L^1(\mathbf{R})$ is defined to be the set of all integrable functions on $\mathbf{R}$. We define

$$\|f\|_1 = \int |f|. \tag{1}$$

Then $\|f\|_1 \geq 0$ and we know that $\|f\|_1 = 0$ if and only if $f = 0$ a.e. Therefore, $\| \|_1$ is a *norm* on the vector space L^1 only if we *identify*, that is, do not distinguish between, functions that are equal a.e. The remaining properties of a norm are easily established:

$$\|af\|_1 = |a| \cdot \|f\|_1, \quad a \in \mathbf{R}; \qquad \|f + g\|_1 \leq \|f\|_1 + \|g\|_1.$$

Thus, up to identifying functions that differ only on a null set, L^1 has a *metric*

$$d(f, g) = \|f - g\|_1.$$

As a metric space, L^1 has two properties of great importance: It is *complete* (every Cauchy sequence converges) and the integrable *continuous* functions are a dense subset.

Lemma 12.4. *Suppose that h is a nonnegative measurable function. For any $\lambda > 0$,*

$$m(\{h > \lambda\}) \leq \frac{1}{\lambda} \int h.$$

Proof: Let $A = \{h > \lambda\}$. Then $h > \lambda 1_A$, so $\int h \geq \int \lambda 1_A = \lambda \, m(A)$. (We have used this idea several times already.) $\square$

Theorem 12.5. *L^1 is complete: Given a Cauchy sequence $(f_n)_{n=1}^\infty$ in L^1, there is a function $f \in L^1$ such that*

$$\lim_{n \to \infty} \|f_n - f\|_1 = 0.$$

Proof: Given a positive integer k, choose $N(k)$ so large that $\|f_n - f_m\|_1 < 2^{-k}$ if $n, m \geq N(k)$. Also choose these integers so that $N(k+1) > N(k)$. It is enough to show that the subsequence $g_k = f_{N(k)}$ has a limit. Note that

$$\|g_k - g_{k+1}\|_1 \leq 2^{-k}.$$

Set $h_1(x) = |g_1(x)|$ and

$$h_{n+1}(x) = |g_1(x)| + \sum_{k=1}^{n} |g_{k+1}(x) - g_k(x)|.$$

Then $(h_n)_{n=1}^{\infty}$ is nonnegative and nondecreasing and

$$\int h_n \leq \|g_1\|_1 + \sum_{k=1}^{n} \|g_{k+1} - g_k\|_1 \leq \|g_1\|_1 + \sum_{k=1}^{n} 2^{-k} = \|g_1\|_1 + 1.$$

By the Monotone Convergence Theorem, the limit

$$|g_1(x)| + \sum_{n=1}^{\infty} |g_{n+1}(x) - g_n(x)|$$

is integrable; therefore it is finite a.e. Writing

$$g_n(x) = g_1(x) + \sum_{k=1}^{n} \left[g_{k+1}(x) - g_k(x)\right],$$

we see that $f(x) = \lim g_n(x)$ exists (it is the sum of an absolutely convergent series) for a.e. x. Moreover,

$$|f(x) - g_n(x)| \leq \sum_{k=n}^{\infty} |g_{k+1}(x) - g_k(x)|,$$

so

$$\|f - g_n\|_1 \leq \sum_{k=n}^{\infty} \|g_{k+1} - g_k\|_1 \leq 2^{1-n}$$

and f is the desired limit. □

An example. Let f_0 be the indicator function of the interval $[0, 1]$, and f_1, f_2, $f_3, \ldots$, the indicator functions of the successive intervals

$$[0, \tfrac{1}{2}], [\tfrac{1}{2}, 1], [0, \tfrac{1}{3}], [\tfrac{1}{3}, \tfrac{2}{3}], [\tfrac{2}{3}, 1], [0, \tfrac{1}{4}], [\tfrac{1}{4}, \tfrac{1}{2}], [\tfrac{1}{2}, \tfrac{3}{4}], \ldots.$$

Then $\|f_n\|_1 \to 0$. But for any $x \in [0, 1]$ there are infinitely many indices n such that $f_n(x) = 1$. Thus the passage to a *subsequence* is necessary to obtain pointwise convergence. (In this case one could take as a subsequence the functions $f_{n(n-1)/2}$, which are the indicator functions of the intervals $[0, \tfrac{1}{n}]$).

Definition. A function $f : \mathbb{R} \to \mathbb{R}$ is said to be *compactly supported* if there is some $M \geq 0$ such that $f(x) = 0$ if $|x| > M$. Note that any compactly supported continuous function is integrable.

Lemma 12.6. *If C is an open subset of $\mathbb{R}$ and B is a closed, bounded, nonempty subset of C, then there is a compactly supported continuous function g such that*

$$0 \leq g \leq 1, \quad g(x) = 1 \quad \text{if } x \in B, \quad g(x) = 0 \quad \text{if } x \in C^c.$$

Proof: The functions $d(x, B) = \inf\{|x - y| : y \in B\}$ and $d(x, C^c)$ are continuous functions of x that vanish precisely on B and on C^c, respectively. We take

$$g_0(x) = \min\{1, \, d(x, A^c)/d(x, B)\},$$

where A is the intersection of C with a bounded open set that contains B. Then g has the desired properties. □

Theorem 12.7. *The compactly supported continuous functions are dense in L^1; that is, for each f in L^1 and each $\varepsilon > 0$, there is a continuous function g such that g has compact support and $\|f - g\|_1 < \varepsilon$.*

Sketch of proof. By the definitions, there are ISFs $f_1^\pm$ such that $\|f^\pm - f_1^\pm\|_1 < \varepsilon/4$. Then, with $f_1 = f_1^+ - f_1^-$ we have $\|f - f_1\|_1 < \varepsilon/2$. By definition, the ISF f_1 is a linear combination of indicator functions of sets having finite measure. Therefore it is enough to prove that we can approximate each such an indicator function 1_A. We use the approximation property for measurable sets (Section 10D): If $m(A) < \infty$, then, given $\varepsilon > 0$, there is a bounded closed set B and an open set C such that $B \subset A \subset C$ and $m(C \setminus B) < \varepsilon$. The corresponding function g of Lemma 12.6 has the property

$$\|g - 1_A\|_1 < \varepsilon. \quad \square$$

Remark. It follows from this theorem that if we started by using only very nice functions – continuous, compactly supported – but then wanted to take all *limits* with respect to the norm $\| \, \|_1$, we would end up with all the Lebesgue measurable integrable functions. This is analogous to starting with the rational numbers and then wanting to include all limits of Cauchy sequences: The result is all real numbers.

It is often convenient, and sometimes necessary, to consider *complex-valued functions*. A function $f : \mathbb{R} \to \mathbb{C}$ can be written uniquely as the sum $f = g + ih$, where g and h are real-valued; in analogy with complex constants, we refer to them as the *real part* and the *imaginary part* of f:

$$f = g + ih, \qquad g = \text{Re}(f), \qquad h = \text{Im}(f).$$

The function f is said to be *measurable* if its real and imaginary parts are measurable. If so, then f is *integrable* if its real and imaginary parts are integrable, and the integral is

$$\int f = \int g + i \int h.$$

Equivalently, a measurable function f is integrable if its complex modulus $|f|$, defined by $|f|(x) = |f(x)|$, is integrable. Again

$$\left| \int f \right| \leq \int |f|.$$

The theorems, and proofs, in this section remain valid if $L^1(\mathbb{R})$ is taken to be the space of integrable complex-valued functions on $\mathbb{R}$.

In addition to considering functions on the line, it is common to consider (real or complex) integrable functions on an interval I, such as $I = [0, 1]$ or $I = (0, \infty)$. The notation is $L^1(I)$. Theorems 12.5 and 12.7 are valid in this context as well.

Exercises

1. Show that if A and B are measurable set with finite measure, then

$$m(A \triangle B) = \int |1_A - 1_B|, \qquad A \triangle B = (A \cap B^c) \cup (A^c \cap B).$$

   Use this and the approximation property for measurable sets to prove that, for any ISF f and any $\varepsilon > 0$, there is a step function g (constant on *intervals*) such that

$$\int |f - g| < \varepsilon.$$

2. Suppose that the real-valued function f on $\mathbb{R}$ is nonnegative and continuous, and suppose that $\int f < \infty$. Is f necessarily bounded? Prove or give a counterexample.

3. Suppose that f is real-valued and integrable. Given a in $\mathbb{R}$, define the translate f_a by $f_a(x) = f(x - a)$. Prove that

$$\lim_{a \to 0} \|f_a - f\|_1 = 0.$$

4. Suppose that f and g are integrable on $\mathbb{R}$ and set $F(x) = \int_a^x f$ and $G(x) = \int_a^x g$. Prove the general integration-by-parts formula

$$\int_a^b Fg = F(b)G(b) - \int_a^b fG.$$

5. Prove a version of *Lusin's Theorem*: If f is integrable, then for each $\varepsilon > 0$ there is a set A with measure $< \varepsilon$ such that f is continuous on the complement of A.
6. Prove that the integrability assumption in Exercise 5 can be replaced by the assumption that f is finite a.e.

12D. The Space L^2

By definition, $L^2 = L^2(\mathbb{R})$ is the set consisting of all real-valued measurable functions f such that the square f^2 is integrable. The elementary inequalities

$$2|ab| \leq a^2 + b^2, \qquad (a+b)^2 \leq 2(a^2 + b^2), \qquad a, b \in \mathbb{R}$$

imply that if f and g belong to L^2, then so does the sum $f + g$, while the product fg is integrable. In particular, L^2 is a vector space, and we may define an *inner product* in L^2:

$$(f, g) = \int fg, \qquad f, g \in L^2. \tag{2}$$

This has the defining properties of an inner product:

$$(f, f) \geq 0; \quad (f, f) = 0 \quad \text{if and only if } f = 0 \text{ a.e.};$$
$$(f, g) = (g, f);$$
$$(af, g) = a(f, g), \qquad a \in \mathbb{R};$$
$$(f + g, h) = (f, h) + (g, h).$$

One can deduce, therefore, the usual *Cauchy-Schwarz inequality*:

$$(f, g)^2 \leq (f, f)(g, g).$$

(See Exercise 5.) The inner product induces a norm and a metric in L^2:

$$\|f\|_2 = (f, f)^{1/2}, \qquad d_2(f, g) = \|f - g\|_2.$$

In particular, the triangle inequality follows from the Cauchy-Schwarz inequality. The Cauchy-Schwarz inequality may be written in terms of the norm:

$$|(f, g)| \leq \|f\|_2 \|g\|_2. \tag{3}$$

The following theorem says that $L^2(\mathbb{R})$, with its inner product as given above, is a (real) *Hilbert space*: a vector space, with an inner product, that is complete with respect to the metric induced by the inner product.

Theorem 12.8. *L^2 is complete.*

12D. The Space L^2

Proof: As in the proof for L^1, we may choose a subsequence $(g_n)_{n=1}^\infty$ of a given Cauchy sequence $(f_n)_{n=1}^\infty$ such that

$$\|g_{n+1} - g_n\|_2 \le 2^{-n}.$$

If A is a bounded interval, then the Cauchy-Schwarz inequality gives

$$\int_A |g_{n+1} - g_n| \le \|g_{n+1} - g_n\|_2 \cdot \|1_A\|_2 = \|g_{n+1} - g_n\|_2 \cdot m(A)^{1/2}.$$

As in the proof of completeness of L^1, it follows that $f(x) = \lim g_n(x)$ exists a.e. in the interval A. But $\mathbb{R}$ is the union of countably many bounded intervals, so the limit f exists a.e. on $\mathbb{R}$.

To show that $\|g_n - f\|_2 \to 0$, we let h be any function in L^2. Then, as before,

$$[f(x) - g_n(x)]h(x) = \sum_{k=n}^\infty [g_{k+1}(x) - g_k(x)]\, h(x);$$

$$|(f - g_n, h)| \le \sum_{k=n}^\infty \int |g_{k+1} - g_k|\, |h| \le \sum_{k=n}^\infty \|g_{k+1} - g_k\|_2 \|h\|_2 \le 2^{1-n} \|h\|_2.$$

Letting $h = f - g_n$, we get

$$\|f - g_n\|_2^2 \le 2^{1-n} \|f - g_n\|_2,$$

so $\|f - g_n\|_2 \le 2^{1-n}$. (Strictly speaking, we should let h run through an appropriate sequence approximating $f - g_n$, since we do not know ahead of time that $f - g_n$ actually is square integrable.) □

Theorem 12.9. *For each function f in L^2 and each $\varepsilon > 0$ there is a compactly supported continuous function g such that $\|f - g\|_2 \le \varepsilon$.*

(See the proof sketched for L^1; it also applies here.)

Again, it is often convenient to consider complex-valued functions. The complex version of $L^2(\mathbb{R})$ consists of functions with a square-integrable modulus: $\int |f|^2 < \infty$. Then the inner product must be modified:

$$(f, g) = \int f \overline{g}.$$

The properties of this inner product are the same as above, except that

$$(f, g) = \overline{(g, f)}.$$

The Cauchy-Schwarz inequality remains valid, and the proofs of the two theorems carry over, so complex L^2 is a complex Hilbert space.

Again, one frequently encounters spaces $L^2(I)$ of functions square integrable on a real interval I. Theorems 12.8 and 12.9 remain valid.

Exercises

In these exercises, the norm, inner product, and convergence notions are those of $L^2(\mathbb{R})$.

1. Suppose that $\lim_{n\to\infty} f_n = f$. Prove that $\lim_{n\to\infty}(f_n, g) = (f, g)$.
2. Suppose that $\lim_{n\to\infty} f_n = f$ and $\lim_{n\to\infty} g_n = g$. Prove that $\lim_{n\to\infty}(f_n, g_n) = (f, g)$.
3. Show that if f and h are *orthogonal*, that is, $(f, h) = 0$, then

$$||f + h||^2 = ||f||^2 + ||h||^2.$$

4. Suppose that $f \neq 0$. Show that for each g there is a unique constant a such that $g - af$ is orthogonal to f.
5. Prove the Cauchy-Schwarz inequality $|(f, g)| \leq ||f|| \cdot ||g||$ by writing $g = af + h$ as in the previous exercise, with $h \perp f$ and noting that $||g||^2 = ||af||^2 + ||h||^2 \geq ||af||^2$. (Note that neither of the preceding two exercises depends on the Cauchy-Schwarz inequality, so the reasoning is not circular.)
6. Suppose that $f_1, f_2, \ldots, f_n$ are mutually orthogonal: $(f_j, f_k) = 0$ if $j \neq k$. Prove that

$$||f_1 + f_2 + \cdots + f_n||^2 = ||f_1||^2 + ||f_2||^2 + \cdots ||f_n||^2.$$

7. Suppose that the functions $e_1, e_2, \ldots, e_n$ are *orthonormal*, that is, $(e_j, e_j) = 1$, and $(e_j, e_k) = 0$ if $j \neq k$. Show that for each f there is a unique linear combination $g = \sum a_j e_j$ such that $f - g$ is orthogonal to each of the e_k's.
8. In the preceding exercise, show that $\sum |a_n|^2 \leq ||f||^2$. When does equality hold?
9. Show that the element g in Exercise 7 is the closest element to f in the subspace spanned by the elements $e_1, \ldots, e_n$.
10. Suppose that f is orthogonal to each element of a subset S. Show that f is orthogonal to each element in the closure of the subspace spanned by S.
11. Suppose that f is real-valued and belongs to L^2. Set

$$F(x) = \int_0^x f(t)\, dt, \qquad x \in \mathbb{R}.$$

Prove that there is a constant C such that $|F(y) - F(x)| \leq C|y - x|^{1/2}$, all $x, y \in \mathbb{R}$.

12. Tie up the loose end in the proof of completeness in L^2: Suppose that f is a measurable real-valued function on $\mathbb{R}$ and suppose that for every h in L^2 the function fh is integrable with $|\int fh| \leq C||h||$. Prove that $||f|| \leq C$.

12E. Differentiating the Integral

Suppose that f is an integrable function on $\mathbb{R}$. Fixing a point $a \in \mathbb{R}$, we may define

$$F(x) = \int_{[a,x]} f, \quad \text{if } x \geq a; \qquad F(x) = -\int_{[x,a]} f, \quad \text{if } x < a.$$

By definition, we write this as a single formula:

$$F(x) = \int_a^x f.$$

Theorem 12.10: Differentiation. *For a.e. x, the derivative $F'(x)$ exists and equals $f(x)$.*

The proof is not trivial; we outline the steps.

First, F is continuous. In fact, this is an easy consequence of the Dominated Convergence Theorem applied to a sequence of functions $f 1_{A_n}$, where $A_n = [a, x_n]$ or $A_n = [x_n, a]$, where $\lim x_n = x$.

Second, the result is true if f is continuous: This is just the ordinary differentiation theorem.

Third, we may approximate f by continuous functions, since they are dense in L^1. However, in order to control what happens when we approximate, we need some machinery.

Definition. The Hardy-Littlewood *maximal function* of a function $h \in L^1$ is defined by

$$h^*(x) = \sup \left\{ \frac{1}{|I|} \int_I |h| : x \in I, \ I \text{ an open interval} \right\}.$$

Theorem 12.11: The Hardy-Littlewood inequality. *For any positive λ, the set $E_\lambda = \{h^* > \lambda\}$ has measure*

$$m(E_\lambda) \le \frac{5}{\lambda} \int |h|. \tag{4}$$

Proof of Theorem 12.10, assuming Theorem 12.11. We want to show that the the set where

$$\limsup_{y \to x} \left| \frac{F(y) - F(x)}{y - x} - f(x) \right| > 0$$

has measure 0. It is enough to show that for each $\delta > 0$ and $\varepsilon > 0$ the set

$$E = \left\{ x : \limsup_{y \to x} \left| \frac{F(y) - F(x)}{y - x} - f(x) \right| > \delta \right\}$$

has measure $\le M\varepsilon$ for some fixed M, because then we may take a sequence δ_n with limit 0. Choose a continuous $g \in L^1$ such that $\|f - g\|_1 < \varepsilon\delta$. Set $G(x) = \int_a^x g$.

Given $y \neq x$, let I be the open interval with endpoints x, y. Then

$$\left| \frac{F(y) - F(x)}{y - x} - f(x) \right| \leq \left| \frac{F(y) - F(x)}{y - x} - \frac{G(y) - G(x)}{y - x} \right|$$
$$+ \left| \frac{G(y) - G(x)}{y - x} - g(x) \right| + |g(x) - f(x)|$$
$$= \frac{1}{|I|} \left| \int_I (f - g) \right| + \left| \frac{G(y) - G(x)}{y - x} - g(x) \right|$$
$$+ |g(x) - f(x)|.$$

The middle term in the last expression goes to zero as $y \to x$. The first and last terms are at most $h^*(x)$ and $|h(x)|$, respectively, where $h = f - g$. Therefore E is contained in the union of the two sets

$$\{h^* > \delta\}, \quad \{h > \delta\}.$$

The measures of these sets are, respectively, at most

$$5 \cdot \frac{2}{\delta} \int |h| \leq 10\varepsilon, \quad \frac{2}{\delta} \int |h| \leq 2\varepsilon.$$

Thus $m(E) \leq 12\varepsilon$ and the proof is complete. □

We turn now to the Hardy-Littlewood maximal inequality (4) and assume again that h is in L^1 and that $\lambda > 0$ is given. Again

$$E_\lambda = \{x : h^*(x) > \lambda\}.$$

Now a point x belongs to E_λ if and only if there is an open interval I with $x \in I$ such that

$$\frac{1}{|I|} \int_I |h| > \lambda.$$

Thus $|I| < \int_I |h|/\lambda$, so there is a bound to the lengths of intervals I. Notice also that $I \subset E_\lambda$. Thus, if $I_1, I_2, \ldots$ is a pairwise disjoint family of such intervals with union $A \subset E_\lambda$, then

$$m(A) = \sum |I_n| \leq \frac{1}{\lambda} \sum \int_{I_n} |h| = \frac{1}{\lambda} \int_A |h| \leq \frac{1}{\lambda} \int |h|.$$

We can finish the job with a *covering lemma*.

Lemma 12.12. *Suppose that $E \subset \mathbb{R}$ is the union of a family $\mathcal{I}$ of open intervals of length $\leq c < +\infty$. Then there is a finite or countable collection of pairwise disjoint* intervals $I_1, I_2, \ldots$ belonging to the family $\mathcal{I}$ such that

$$m(E) \leq 5(|I_1| + |I_2| + \cdots).$$

Proof: Let c_1 be the supremum of the lengths of the intervals in $\mathcal{I}$ and choose $I_1 \in \mathcal{I}$ such that $|I_1| \geq c_1/2$. Suppose now that we have chosen intervals $I_1, \ldots, I_n$ and define a subfamily $\mathcal{I}_n$ consisting of those intervals in $\mathcal{I}$ that are disjoint from each of $I_1, \ldots, I_n$. If $\mathcal{I}_n$ is empty, stop. Otherwise, choose $I_{n+1} \in \mathcal{I}_{n+1}$ such that

$$|I_{n+1}| \geq \tfrac{1}{2} c_{n+1}, \qquad c_{n+1} = \sup\{|I| : I \in \mathcal{I}_{n+1}\}.$$

Case 1. For some n, the collection $\mathcal{I}_{n+1}$ is empty. Then every interval $I \in \mathcal{I}$ intersects one of $I_1, \ldots, I_n$. Suppose that k is the first index such that $I \cap I_k \neq \emptyset$. Because of the method of choice, this implies that $|I_k| \geq |I|/3$, so that $I \subset I_k^*$, which is the interval with the same midpoint as I_k but five times the length. This shows that

$$E \subset \bigcup_{k=1}^{n} I_k^*,$$

and the result follows.

Case 2. We can choose I_n for all n. If $\sum_{n=1}^{\infty} |I_n| = \infty$, then there is nothing to prove. Otherwise, $|I_n| \to 0$ as $n \to \infty$. Given $\emptyset \neq I \in \mathcal{I}$, we have $|I_n| \leq |I|/2$ for large n. Considering how the intervals I_n were chosen, this implies that I intersects some I_n, and, as in Case 1, this implies that $I \subset I^*$ for some k. Thus again

$$E \subset \sum_{n=1}^{\infty} I_n^*. \qquad \square$$

Exercises

1. Let $f(x) = 0$ outside the interval $(0, \tfrac{1}{2})$ and $f(x) = 1/x(\log x)^2$ for $x \in (0, \tfrac{1}{2})$. Show that f is in $L^1(\mathbb{R})$ but the maximal function f^* is not.
2. Suppose that f is in $L^1(\mathbb{R})$. Prove that $f^* \geq |f(x)|$ a.e.
3. Suppose that A is measurable. A point $x \in A$ is said to be a *point of density* for A if

$$\lim_{|I| \to 0} \frac{m(A \cap I)}{m(I)} = 1,$$

where the limit is taken over intervals I that contain x. Prove that almost every point of A is a point of density for A.

Additional Exercises for Chapter 12

These exercises deal with L^p spaces for values of p more general than $p = 1, p = 2$. For $0 < p < \infty$, the space $L^p = L^p(\mathbb{R})$ consists of all measurable functions f such that $|f|^p$ is integrable. Let

$$\|f\|_p = \left(\int |f|^p\right)^{1/p}.$$

1. Suppose that p and q are positive and $1/p + 1/q = 1$, so $p > 1$ and $q > 1$.
   (a) Prove that for any nonnegative a and b,
   $$ab \leq \frac{a^p}{p} + \frac{b^q}{q}.$$
   (b) Prove that if f belongs to L^p and g belongs to L^q, then the product fg belongs to L^1 and
   $$\left|\int fg\right| \leq \frac{1}{p}\int |f|^p + \frac{1}{q}\int |g|^q.$$

2. Prove *Hölder's inequality*: If f is in L^p and g is in L^q, and $1/p + 1/q = 1$, then
   $$\left|\int fg\right| \leq \|f\|_p \|g\|_q.$$
   (Note that the Cauchy-Schwarz inequality (3) is a special case.)

3. For $1 < p < \infty$, prove that the result in Exercise 2 is optimal: If $1/p + 1/q = 1$, then
   $$\|f\|_p = \sup\left\{\left|\int fg\right| : \|g\|_q = 1\right\}.$$

4. Show that the expression $\|f\|_p$ has the properties of a norm (if we identify functions that coincide a.e.) when $p \geq 1$. In particular, prove the triangle inequality
   $$\|f + g\|_p \leq \|f\|_p + \|g\|_p$$
   for $1 < p < \infty$.

5. Suppose that $0 < p < 1$. Prove that the triangle inequality fails: There are f, g in L^p such that
   $$\|f + g\|_p > \|f\|_p + \|g\|_p.$$

6. Show that for each p, $0 < p < \infty$, there is a function f that belongs to L^p but does not belong to any L^q for $q \neq p$.

13
Fourier Series

The subject of Fourier series has its roots in the analysis of certain problems with a physical origin, including vibrations of a string and conduction of heat, and it has many mathematical applications and ramifications as well. In this chapter we introduce the basic question – how to represent a general periodic function in terms of simpler ones – and develop the fundamental mathematical ideas associated to this process.

13A. Periodic Functions and Fourier Expansions

A real- or complex-valued function f defined on the line is said to be *periodic*, with *period* $L > 0$, if $f(x + L) = f(x)$ for every x. If so, then it is also periodic with period nL for every positive integer n. The most familiar examples are the trigonometric functions $\sin x$ and $\cos x$, with period 2π. We can always rescale a function g that has period L to a function f that has period 2π by setting $f(x) = g(xL/2\pi)$. Throughout this chapter, we consider functions with period 2π and refer to such functions simply as "periodic functions." Depending on the context, we may only require functions f to be defined almost everywhere and to satisfy the basic equation

$$f(x + 2\pi) = f(x)$$

only for almost every x.

The following is an elementary but important observation. Suppose that I is a half-open interval of length 2π, say $I = [a, a + 2\pi)$, and suppose that f is defined on I. Then there is a unique extension of f that is defined on the entire line and is periodic. In fact, for any integer n, a point x belongs to the interval $[a + 2n\pi, a + 2(n + 1)\pi)$ if and only if $x - 2n\pi$ belongs to I, so we must set $f(x) = f(x - 2n\pi)$. Usually we take $I = [-\pi, \pi)$ and simply identify functions on I and periodic functions on the line.

The simplest continuous periodic functions are the trigonometric functions $\sin nx$ and $\cos nx$, indexed by the positive integers and the nonnegative integers, respectively. It turns out to be more convenient to consider instead the complex exponential functions

$$e_n(x) = e^{inx} = \sum_{m=0}^{\infty} \frac{(inx)^m}{m!} = \cos nx + i \sin nx, \quad n \in \mathbb{Z}. \tag{1}$$

A linear combination of periodic functions is a periodic function, and a function that is, at each point, the limit of periodic functions is also a periodic function. Therefore, any two-sided series

$$\sum_{-\infty}^{\infty} a_n e^{inx} = \lim_{N \to \infty} \sum_{n=-N}^{N} a_n e^{inx} \tag{2}$$

that converges at every point defines a periodic function. (Here $\{a_n\}_{-\infty}^{\infty}$ is a two-sided sequence of complex constants.)

It is easy to produce examples. In fact, each of the exponentials has modulus 1 at each point, so if the sum $\sum_{-\infty}^{\infty} |a_n|$ is finite,

$$\sum_{n=-\infty}^{\infty} |a_n| = \sum_{n=1}^{\infty} |a_n| + \sum_{n=0}^{\infty} |a_{-n}| < \infty, \tag{3}$$

then the series (2) converges absolutely at every point; in fact, it converges uniformly, so the limit is a continuous function (see Theorem 7.8).

The fundamental questions to be investigated in this chapter are

- Given a periodic function, can it be expressed as a series (2)?
- If so, is the expression unique?
- How can the coefficients a_n be calculated?
- How are properties of f related to properties of the coefficients a_n?

Here is an argument that suggests that, in some sense, the answers to the first two questions should be "yes." The argument also may help to explain the special role of the exponentials (1) and shows how to calculate the a_n's.

We start by considering a finite-dimensional complex vector space V with inner product ( ). A basis $\mathbf{e}_1, \ldots, \mathbf{e}_n$ for V is said to be *orthonormal* if the basis vectors are unit vectors that are mutually orthogonal:

$$(\mathbf{e}_j, \mathbf{e}_j) = 1, \quad (\mathbf{e}_j, \mathbf{e}_k) = 0 \quad \text{if } j \neq k. \tag{4}$$

Any element $\mathbf{v} \in V$ can be written uniquely as a linear combination of the basis vectors $\mathbf{v} = \sum a_k \mathbf{e}_k$. Moreover, if we take the inner product of each side of this

13A. Periodic Functions and Fourier Expansions

equality with $\mathbf{e}_j$ and use (4), we can identify $(\mathbf{v}, \mathbf{e}_j) = a_j$. Therefore

$$\mathbf{v} = \sum_{j=1}^{n} (\mathbf{v}, \mathbf{e}_j) \mathbf{e}_j. \tag{5}$$

There are various ways to *find* an orthonormal basis for a given finite-dimensional inner product space V. One way uses the *spectral theorem*: Suppose that $T : V \to V$ is a linear transformation that satisfies the symmetry condition

$$(T\mathbf{v}, \mathbf{w}) = (\mathbf{v}, T\mathbf{w}), \quad \text{all } \mathbf{v}, \mathbf{w} \in V. \tag{6}$$

Then there is an orthonormal basis for V that consists of eigenvectors of T: $T\mathbf{e_j} = \lambda_j \mathbf{e}_j$ for some constants λ_j.

We know from Chapter 12 how to find an inner product space that includes a large class of functions on the interval $I = [-\pi, \pi)$; as we noted above, such functions can be identified with periodic functions. Thus, by the space of L^2-*periodic functions*, or *square-integrable periodic functions*, we mean the periodic functions identified with the functions that are square integrable on the basic interval, that is, functions in the space

$$L^2(I), \quad I = [-\pi, \pi). \tag{7}$$

We take the inner product in this space to be given by

$$(f, g) = \frac{1}{2\pi} \int_{-\pi}^{\pi} f(x) \overline{g(x)} \, dx.$$

We define an operator T for periodic functions f that have continuous derivatives by

$$Tf(x) = -if'(x). \tag{8}$$

It is the factor $-i$ that makes this operator symmetric. In fact, suppose that f and g are two such functions. Integration by parts shows that

$$(Tf, g) = -i \frac{1}{2\pi} \int_{-\pi}^{\pi} f'(x) \overline{g(x)} \, dx$$

$$= -if(x) g(x) \Big|_{-\pi}^{\pi} + i \frac{1}{2\pi} \int_{-\pi}^{\pi} f(x) \overline{g'(x)} \, dx$$

$$= \frac{1}{2\pi} \int_{-\pi}^{\pi} f(x) \overline{[-ig'(x)]} \, dx = (f, Tg).$$

(The boundary terms cancel because of periodicity.)

This suggests, but is far from a proof, that the eigenvectors of T might be a basis for our space (6). An eigenvector would be a function that is a solution of $-if'(x) = \lambda f(x)$ and therefore a multiple of

$$f(x) = e^{i\lambda x}.$$

This function is periodic only if $f(2\pi) = e^{i2\lambda\pi} = f(0) = 1$, which is true if and only if λ is an integer (Section 9A, Exercise 1). Thus the eigenvectors of T are, up to multiplication by constants, the exponentials (1). Let us check the orthonormal conditions. It will help to recall various facts about the exponential function with an imaginary argument:

$$e^{ix} e^{iy} = e^{i(x+y)}; \qquad \overline{e^{ix}} = e^{-ix}; \qquad |e^{ix}| = 1.$$

Then

$$(e_n, e_n) = \frac{1}{2\pi} \int_{-\pi}^{\pi} |e_n(x)|^2 \, dx = \frac{1}{2\pi} \int_{-\pi}^{\pi} 1 \, dx = 1;$$

$$(e_n, e_m) = \frac{1}{2\pi} \int_{-\pi}^{\pi} e^{inx} e^{-imx} \, dx = \frac{1}{2\pi} \int_{-\pi}^{\pi} e^{i(n-m)x} \, dx$$

$$= \frac{1}{2\pi} \int_{-\pi}^{\pi} \frac{d}{dx} \left\{ \frac{e^{i(n-m)x}}{i(n-m)} \right\} dx = \left. \frac{e^{i(n-m)x}}{i(n-m)} \right|_{-\pi}^{\pi} = 0, \qquad n \neq m. \quad (9)$$

This discussion suggests that we try to write a periodic function f as a series by the formal recipe

$$f(x) = \sum_{n=-\infty}^{\infty} \widehat{f}(n) e^{inx}, \tag{10}$$

where the coefficients are

$$\widehat{f}(n) = (f, e_n) = \frac{1}{2\pi} \int_{-\pi}^{\pi} f(y) e^{-iny} \, dy, \qquad n \in \mathbb{Z}. \tag{11}$$

(We use y as the variable of integration here to distinguish it from the variable x in (10).)

Expressions equivalent to (10 and (11) were first proposed by Fourier in his study of heat flow and are now called the *Fourier series* and *Fourier coefficients* of the function f. In the rest of this chapter we study the validity of (10).

13B. Fourier Coefficients of Integrable and Square-Integrable Periodic Functions

The necessary and sufficient condition for all the integrals (11) to exist is that the function f be integrable on the period interval $I = [-\pi.\pi)$. We say that a measurable complex-valued function f on $\mathbb{R}$ is an *integrable periodic function* if it satisfies $f(x + 2\pi) = f(x)$ a.e. and its restriction to I is integrable.

Definition. The *Fourier coefficients* of an integrable periodic function f are the complex numbers $\widehat{f}(n)$ defined by the integrals (11).

The L^1 norm of an integrable periodic function will be taken to be

$$\|f\|_1 = \frac{1}{2\pi} \int_{-\pi}^{\pi} |f(x)|\, dx. \tag{12}$$

Similarly, the L^2 norm of a square-integrable periodic function will be determined from

$$\|f\|^2 = (f, f) = \frac{1}{2\pi} \int_{-\pi}^{\pi} |f(x)|^2\, dx. \tag{13}$$

(These conflict with the notation in Chapter 12, but we can live with the conflict.) The relation between these function spaces and norms is as follows.

Lemma 13.1. *An L^2-periodic function is an integrable periodic function, and*

$$\|f\|_1 \leq \|f\|. \tag{14}$$

Proof: The function $|f|$ is also L^2-periodic. We take its inner product with the constant function e_0 and use the Cauchy-Schwarz inequality to obtain

$$\|f\|_1 = (|f|, e_0) \leq \|f\| \cdot \|e_0\| = \|f\|. \qquad \square$$

The propositions in this section list some of the most important properties of the Fourier coefficients of a function.

Proposition 13.2. *Suppose that f is an integrable periodic function. Then*

$$|\widehat{f}(n)| \leq \|f\|_1, \quad \text{all } n \in \mathbb{Z}. \tag{15}$$

Proof: Inequality (15) is immediate from the definition of $\widehat{f}(n)$ and the fact that $|e_n| \equiv 1$. $\qquad \square$

Proposition 13.3. *Suppose that f is an integrable periodic function. Given a in $\mathbb{R}$, let f_a be the translate: $f_a(x) = f(x - a)$. Then*

$$\widehat{f_a}(n) = e^{-ina}\, \widehat{f}(n), \quad \text{for all } n \in \mathbb{Z}. \tag{16}$$

Proof: The identity (16) is left as an exercise. The proof uses the fact that, for a periodic function, the integral over any interval of length 2π is the same:

$$\int_b^{b+2\pi} g(x)dx = \frac{1}{2\pi}\int_{-\pi}^{\pi} g(x)dx \qquad \text{if } g \text{ is periodic.} \tag{17}$$

This identity is also left as an exercise. □

Proposition 13.4. *Suppose that f is a continuous periodic function with continuous derivative f'. Then*

$$\widehat{f'}(n) = in\widehat{f}(n), \qquad \text{all } n \in \mathbb{Z}. \tag{18}$$

Proof: The identity (18) is left as an exercise. □

Proposition 13.5. The Riemann-Lebesgue Lemma. *If f is an integrable periodic function, then*

$$\lim_{|n|\to\infty} \widehat{f}(n) = 0. \tag{19}$$

Proof: It is enough to prove the result for continuous functions, since for any $\varepsilon > 0$ we can choose a continuous periodic g with $\|f - g\|_1 < \varepsilon$. By Proposition 13.2 (and the additivity of integration), the Fourier coefficients of f and g differ by less than ε, so the $\widehat{f}(n)$'s are eventually less than ε in modulus if the $\widehat{g}(n)$ approach 0.

If g is continuous, we may use Proposition 13.3 to estimate $\widehat{g}(n)$ for $n \neq 0$. In fact, choose $a = \pi/n$, so $e^{-ina} = -1$. Then

$$|2\widehat{g}(n)| = |\widehat{g}(n) - \widehat{g}_a(n)| \leq \frac{1}{2\pi}\int_{-\pi}^{\pi}|g(x) - g(x-a)|\,dx, \qquad a = a(n) = \frac{\pi}{n}.$$

As $|n|$ tends to ∞, the corresponding $a = a(n)$ tends to 0, so the last integrand approaches 0 uniformly. □

To visualize the reason that the Fourier coefficients of a continuous function approach zero, look at the real and imaginary parts of e^{inx}, $\cos nx$ and $\sin nx$. These functions oscillate more and more rapidly. We multiply by a fixed continuous function and integrate; since the function is continuous, it is nearly constant over adjacent pairs of intervals of length $\pi/|n|$ while each of $\cos nx$ and $\sin nx$ have opposite signs on such adjacent intervals. Therefore, the integrals over two adjacent intervals nearly cancel. It is plausible – and true – that the sum of all the integrals over subintervals is itself small when $|n|$ is large.

For the proof of the final result in this section, we introduce the partial sums of the Fourier series. If f is an integrable periodic function and N is a nonnegative integer, we define the function $S_N f$ as a partial sum of the Fourier series:

$$S_N f(x) = \sum_{n=-N}^{N} \widehat{f}(n) e^{inx}. \tag{20}$$

Proposition 13.6. Bessel's inequality. *If f is an L^2-periodic function, then*

$$\sum_{-\infty}^{+\infty} |\widehat{f}(n)|^2 \leq \|f\|^2. \tag{21}$$

(We shall see later that there is actually equality in (21).)

Proof: Let $g = f - S_N f$, so that $f = S_N f + g$. A direct computation of the inner products shows that $(g, e_n) = 0$, for $|n| \leq N$. It follows from this that $(g, S_N f) = 0 = (S_N f, g)$. Therefore, using the properties of the inner product, we have

$$\begin{aligned}
\|f\|^2 &= \|S_N f + g\|^2 = (S_N f + g, S_N f + g) \\
&= \|S_N f\|^2 + (S_N f, g) + (g, S_N f) + \|g\|^2 \\
&= \|S_N f\|^2 + \|g\|^2 \geq \|S_N f\|^2.
\end{aligned} \tag{22}$$

In the same way, the orthonormality of the e_n implies that

$$\|S_N f\|^2 = \sum_{-N}^{N} |\widehat{f}(n)|^2. \tag{23}$$

Taken together, (22) and (23) imply that the partial sums of the series in (21) are bounded by $\|f\|^2$. $\square$

Remark. It is worth noting explicitly one consequence of (21) or (22):

$$\|S_N f\| \leq \|f\|. \tag{24}$$

Exercises

1. Compute the Fourier coefficients of the function f when it has the following forms on the interval $(-\pi, \pi)$:

    (a) $\quad f(x) = x;$
    (b) $\quad f(x) = |x|;$
    (c) $\quad f(x) = x^2;$
    (d) $\quad f(x) = \begin{cases} -1, & x \in [-\pi, 0) \\ 1, & x \in [0, \pi). \end{cases}$

2. Suppose that f is a finite sum $\sum_{-N}^{n} c_n e^{inx}$.
   (a) Prove that $\widehat{f}(n) = c_n$.
   (b) Show that f can be written in the form
   $$f(x) = \frac{1}{2}a_0 + \sum_{n=1}^{N}[a_n \cos nx + b_n \sin nx].$$
   (c) Relate a_n, b_n to c_n and also express them as integrals.
   (d) Show that f is real if and only if all a_n, b_n are real.
   (e) Show that f is *even*, that is, $f(-x) = f(x)$, if and only if all $b_n = 0$.
   (f) Show that f is *odd*, that is, $f(-x) = -f(x)$, if and only if all $a_n = 0$.
3. Suppose that the sequence of functions $f_n = \sum_{k=-n}^{n} a_n e^{inx}$ converges in $L^1(-\pi, \pi)$ to the integrable periodic function f. Show that
   $$|n| \geq |m| \Rightarrow \widehat{f}_n(m) = \widehat{f}(m).$$
4. Find a function f whose Fourier coefficients are
   $$\widehat{f}(n) = 0, \quad n < 0; \quad \widehat{f}(n) = 2^{-n}, \quad n > 0.$$
   Be as explicit as possible.
5. Suppose that a is complex and $|a| < 1$. Find the Fourier coefficients of the functions
   (a) $$\frac{1}{1 - ae^{ix}},$$
   (b) $$\frac{1}{1 - ae^{i2x}},$$
   (c) $$\sum_{n=1}^{\infty} a^n \sin(2^n x).$$
6. (a) Prove the identity (17) for continuous periodic functions.
   (b) Prove (17) for arbitrary integrable periodic f.
7. Prove (16).
8. Prove (18).
9. Suppose that f is periodic and satisfies the strong continuity condition (Hölder condition): For some positive constants C and α with $\alpha < 1$,
   $$|f(x+h) - f(x)| \leq C|h|^\alpha, \quad \text{for all } x, y.$$
   Show that the Fourier coefficients satisfy, for some constant C',
   $$|\widehat{f}(n)| \leq C'|n|^{-\alpha}, \quad n = \pm 1, \pm 2, \ldots.$$

13C. Dirichlet's Theorem

The most straightforward question about representing a function f as the sum of its Fourier series is, when is the representation (10) valid at a given point x? This

13C. Dirichlet's Theorem

is a surprisingly difficult question. There are continuous periodic functions whose Fourier series diverge at some point. In 1926, Kolmogorov gave an example of an integrable periodic function whose Fourier series diverges a.e. On the other hand, if f is periodic and has a continuous derivative, then (10) *is* valid at every point. This is a consequence of Dirichlet's Theorem, which is proved in this section.

As a first step, we combine equation (20) for $S_N f$ with the integral expression for the Fourier coefficients, in order to write $S_N f$ as an integral. On the way we use the additivity of the integral and the addition formula for the exponential function:

$$S_N f(x) = \sum_{n=-N}^{N} \widehat{f}(n) e^{inx}$$

$$= \sum_{n=-N}^{N} \frac{e^{inx}}{2\pi} \int_{-\pi}^{\pi} f(y) e^{-iny} \, dy$$

$$= \frac{1}{2\pi} \int_{-\pi}^{\pi} \left\{ \sum_{n=-N}^{N} e^{in(x-y)} \right\} f(y) \, dy. \tag{25}$$

The next step is to evaluate the sum; for convenience, we replace $x - y$ by x. Note that $e^{inx} = (e^{ix})^n$, so that the sum is a geometric progression. As long as $e^{inx} \neq 1$, the sum is

$$\sum_{n=-N}^{N} e^{inx} = e^{-iNx} \sum_{n=0}^{2N} (e^{inx})^n$$

$$= e^{-iNx} \frac{e^{i(2N+1)x} - 1}{e^{ix} - 1}$$

$$= \frac{e^{-ix/2}}{e^{-ix/2}} \cdot \frac{e^{i(N+1)x} - e^{-iNx}}{e^{ix} - 1}$$

$$= \frac{e^{i(N+\frac{1}{2})x} - e^{-i(N+\frac{1}{2})x}}{e^{ix/2} - e^{-ix/2}}$$

$$= \frac{\sin\left(\left[N + \frac{1}{2}\right]x\right)}{\sin \frac{1}{2}x}. \tag{26}$$

(The last step follows from the identity $e^{iy} - e^{-iy} = 2i \sin y$.) See Figure 8.

Proposition 13.7. *Suppose that f is periodic and integrable. Then the partial sum $S_N f$ of its Fourier series is given by the integral*

$$\frac{1}{2\pi} \int_{-\pi}^{\pi} D_N(x - y) f(y) \, dy = \frac{1}{2\pi} \int_{-\pi}^{\pi} D_N(y) f(x - y) \, dy, \tag{27}$$

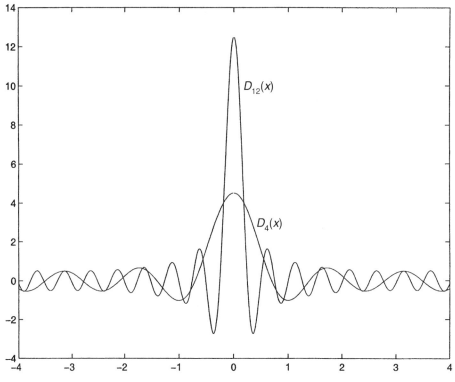

Figure 8. Dirichlet kernel for $N = 4$ and $N = 12$.

where D_N is the **Dirichlet kernel**

$$D_N(x) = \frac{\sin\left([N + \tfrac{1}{2}]x\right)}{\sin \tfrac{1}{2}x} = \sum_{n=-N}^{N} e^{inx}. \qquad (28)$$

(The terminology comes from integral equations. The transformation that takes a function f to a new function Tf defined by

$$Tf(x) = \int K(x, y) f(y) \, dy$$

is said to be an integral transformation with kernel K.)

Proof: The preceding arguments show that $S_N f$ is given by the integral on the left in (28). The fact that the two integrals are equal follows from a change of variable in the integral (take $y = x - y'$) together with (17) to return to the original interval I. □

13C. Dirichlet's Theorem

One basic property of the Dirichlet kernel is

$$\frac{1}{2\pi} \int_{-\pi}^{\pi} D_N(y)\,dy = 1. \tag{29}$$

In fact the integral here is the inner product $(D_N, 1)$; write D_N as a sum of exponentials to see that (29) follows from the orthonormality properties (9).

Theorem 13.8: Dirichlet's Theorem. *Suppose that f is an integrable periodic function that is differentiable at a point x_0. Then*

$$\lim_{N \to \infty} S_N f(x_0) = f(x_0). \tag{30}$$

Proof: We use (29) to rewrite $S_N(x_0) - f(x_0)$ as an integral:

$$S_N(x_0) - f(x_0) = \frac{1}{2\pi} \int_{-\pi}^{\pi} D_N(y) f(x_0 - y)\,dy - \frac{f(x_0)}{2\pi} \int_{-\pi}^{\pi} D_N(y)\,dy$$

$$= \frac{1}{2\pi} \int_{-\pi}^{\pi} D_N(y)[f(x_0 - y) - f(x_0)]$$

$$= \frac{1}{2\pi} \int_{-\pi}^{\pi} \sin\left(\left[N + \tfrac{1}{2}\right]x\right) g(y)\,dy, \tag{31}$$

where

$$g(y) = \frac{f(x_0 - y) - f(x_0)}{y} \cdot \frac{y}{\sin \tfrac{1}{2} y}.$$

By the assumptions on f, the first factor is bounded near 0 and integrable on the interval I. The second factor is bounded on I. Therefore g is integrable on I. It follows that the difference (31) has limit 0 as N tends to infinity. In fact, we may either adapt the proof of the Riemann-Lebesgue Lemma to the last integral in (31) as it stands, or else extend g to vanish on the rest of the interval $(-2\pi, 2\pi)$, extend the integral to this larger integral, and change variables to obtain

$$\frac{1}{2\pi} \int_{-\pi}^{\pi} \sin[2N+1]y) \frac{g(y/2)}{2}\,dy = \frac{1}{2i}[\widehat{h}(2N+1) - \widehat{h}(-2N-1)], \tag{32}$$

where $h(x) = g(x/2)/2$. Therefore (30) follows from the Riemann-Lebesgue Lemma applied to h. □

Remark. It is clear from the proof that instead of differentiability at x_0, we only need to assume that the difference quotient

$$\frac{f(x) - f(x_0)}{x - x_0}$$

is integrable on I. This allows some cusps, such as for $f(x) = \sqrt{|x|}$ on I, $x_0 = 0$.

Exercises

1. (a) Let $f(x) = |x|$ on $[-\pi, \pi)$. Evaluate the Fourier series at $x = 0$ to find
$$\sum_{n=0}^{\infty} \frac{1}{(2n+1)^2}.$$
   (b) Deduce from this the value of $\sum_{n=1}^{\infty} 1/n^2$.
2. Let $f(x) = x^2$. Evaluate the Fourier series at $x = \pi$ to find $\sum_{n=1}^{\infty} 1/n^2$.
3. Compute the Fourier coefficients of $f(x) = e^{ax}$, assuming that a is positive, and prove the identity
$$1 = \frac{\sinh \pi a}{\pi a}\left(1 + 2\sum_{n=1}^{\infty}(-1)^n \frac{a^2}{n^2 + a^2}\right).$$
4. Compute the Fourier coefficients of $f(x) = e^{iax}$, assuming that $0 < a < 1$, and find a representation of $\sec(\pi a)$ as a sum.

13D. Fejér's Theorem

Although the partial sums $S_N f$ need not converge to f even when f is continuous, the situation is much better if one takes averages instead. According to Exercise 3 of Section 3D, the arithmetic means of a convergent sequence necessarily converge. On the other hand, even a nonconvergent sequence like $x_n = (-1)^n$ can have a convergent sequence of arithmetic means. This leads us to look at the arithmetic means of the Fourier series of an integrable periodic function f:

$$T_N f = \frac{S_0 f + S_1 f + \ldots + S_{N-1} f}{N}. \tag{33}$$

As we did for $S_N f$, we can express this as an integral:

$$T_N f(x) = \frac{1}{2\pi}\int_{-\pi}^{\pi} F_N(x-y) f(y)\, dy = \frac{1}{2\pi}\int_{-\pi}^{\pi} F_N(y) f(x-y)\, dy, \tag{34}$$

where the *Fejér kernel* is

$$F_N(x) = \frac{1}{N}\sum_{n=0}^{N-1} D_n(x)$$
$$= \frac{1}{N \sin \frac{1}{2}x}\sum_{n=0}^{N-1} \sin\left(n+\tfrac{1}{2}\right)x$$
$$= \frac{1}{N \sin \frac{1}{2}x}\sum_{n=0}^{N-1} \frac{e^{i(n+\frac{1}{2})x} - e^{-i(n+\frac{1}{2})x}}{2i}. \tag{35}$$

Once again we can write the sums as geometric progressions and evaluate, for $e^{ix} \neq 1$:

$$\sum_{n=0}^{N-1} \frac{e^{i(n+\frac{1}{2})x} - e^{-i(n+\frac{1}{2})x}}{2i} = \frac{e^{ix/2}}{2i} \sum_{n=0}^{N-1} e^{inx} - \frac{e^{-ix/2}}{2i} \sum_{n=0}^{N-1} e^{-inx}$$

$$= \frac{e^{ix/2}}{2i} \frac{e^{iNx} - 1}{e^{ix}} - \frac{e^{-ix/2}}{2i} \frac{e^{-iNx} - 1}{e^{-ix}}$$

$$= \frac{e^{iNx} - 1}{2i(e^{ix/2} - e^{-ix/2})} - \frac{e^{-iNx} - 1}{2i(e^{-ix/2} - e^{ix/2})}$$

$$= \frac{e^{iNx} - 2 + e^{-iNx}}{(2i)^2 \sin \frac{1}{2}x}$$

$$= \frac{\left(e^{i\frac{N}{2}x} - e^{-i\frac{N}{2}x}\right)^2}{(2i)^2 \sin \frac{1}{2}x}$$

$$= \frac{(2i)^2 \sin^2 \frac{N}{2}x}{(2i)^2 \sin \frac{1}{2}x}. \tag{36}$$

Combining (35) and (36), we obtain

$$F_N(x) = \frac{\sin^2 \frac{N}{2}x}{N \sin^2 \frac{1}{2}x}. \tag{37}$$

See Figure 9.

Proposition 13.9. *The Fejér kernel F_N has the properties*

(i) $\qquad\qquad\qquad F_N(x) \geq 0;$

(ii) $\qquad\qquad \frac{1}{2\pi} \int_{-\pi}^{\pi} F_N(x)\,dx = 1;$

(iii) $\qquad\qquad \lim_{N \to \infty} \int_{\delta < |x| < \pi} F_N(x)\,dx = 0 \quad \text{if} \quad 0 < \delta < \pi.$

Proof: Property (i) is obvious from (37). Property (ii) follows from the corresponding property (29) of the Dirichlet kernel. Finally, since $\sin^2 \frac{1}{2}x$ increases as x goes away from the origin in $(-\pi, \pi)$, it follows that

$$F_N(x) \leq \frac{1}{N \sin^2 \frac{\delta}{2}}, \qquad \delta \leq |x| \leq \pi,$$

and property (iii) follows. □

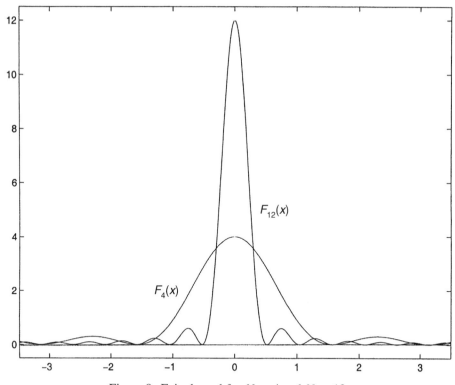

Figure 9. Fejér kernel for $N = 4$ and $N = 12$.

Theorem 13.10: Fejér's Theorem. *If f is continuous and periodic, then the averages $T_N f$ of the partial sums of the Fourier series of f converge uniformly to f.*

Proof: According to Proposition 13.6 and the definition of F_N,

$$T_N f(x) = \frac{1}{2\pi} \int_{-\pi}^{\pi} F_N(y) f(x - y) \, dy. \tag{38}$$

As in the proof of Dirichlet's Theorem, we may take advantage of property (i) in Proposition 13.7 to write

$$T_N f(x) - f(x) = \frac{1}{2\pi} \int_{-\pi}^{\pi} F_N(y) \left[f(x - y) - f(x) \right] dy$$

$$= \frac{1}{2\pi} \int_{|x| < \delta} F_N(y) \left[f(x - y) - f(x) \right] dy$$

$$+ \frac{1}{2\pi} \int_{\delta < |x| < \pi} F_N(y) \left[f(x - y) - f(x) \right] dy \tag{39}$$

for any choice $0 < \delta < \pi$. Properties (i) and (ii) of F_N imply that the first of the two integrals in the last line of (39) has modulus bounded by

$$\frac{1}{2\pi} \cdot \sup\{|f(x-y) - f(x)| : |y| < \delta\}. \tag{40}$$

A continuous periodic function is uniformly continuous (why?), so given $\varepsilon > 0$ we may fix δ so small that the bound (40) is $< \varepsilon/2$, for all N. The modulus of the second integral is bounded by

$$\frac{1}{2\pi} \cdot 2\sup\{|f(y)|\} \cdot \int_{\delta < |y| < \pi} F_N(y)\, dy. \tag{41}$$

Property (3) implies that for all large N, the bound (41) is $< \varepsilon/2$. □

Exercises

1. Suppose that f and g are continuous periodic functions.
   (a) Show that f is even if and only if, for each n, $\widehat{f}(-n) = \widehat{f}(n)$.
   (b) Show that f is odd if and only if, for each n, $\widehat{f}(-n) = -\widehat{f}(n)$.
   (c) Show that $g(x) = \overline{f(x)}$ a.e. if and only if, for each n, $\widehat{g}(n) = \overline{\widehat{f}(-n)}$.
2. Suppose that $f : \mathbb{R} \to \mathbb{R}$ is continuous and periodic. Suppose that

$$\frac{1}{2\pi} \int_{-\pi}^{\pi} f(x) \sin(nx)\, dx = 0 = \frac{1}{2\pi} \int_{-\pi}^{\pi} f(x) \cos(nx)\, dx, \quad n = 0, 1, 2, \ldots.$$

Prove that f is identically zero. (For a considerably more general result, see Exercise 1 of Section 13E.)

13E. The Weierstrass Approximation Theorem

The Weierstrass Approximation Theorem that was proved in Section 7D is one of two results that commonly go by that name. The other refers specifically to the approximation of a continuous *periodic* function by the simplest kind of periodic functions, the trigonometric polynomials.

A *trigonometric polynomial* of degree N is a linear combination of the functions $\{e_n : |n| \le N\}$. The name is justified by the identity $e^{inx} = (\cos x + i \sin x)^n$, which shows that a trigonometric polynomial of degree N is a polynomial of degree N in the functions $\cos x$ and $\sin x$.

Theorem 13.11: The Weierstrass Approximation Theorem. *Any continuous periodic function f can be approximated uniformly by trigonometric polynomials.*

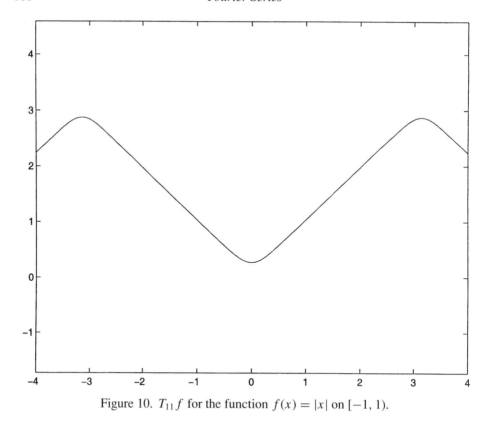

Figure 10. $T_{11}f$ for the function $f(x) = |x|$ on $[-1, 1)$.

Proof: This is an immediate consequence of Fejér's Theorem, because the functions T_N are trigonometric polynomials. □

See Figure 10.

Remarks. 1. Theorem 13.11 predates Theorem 13.10. In fact, Weierstrass recognized that the properties (i), (ii), and (iii) listed in the last section imply uniform convergence, as in the proof of Theorem 13.10. He constructed a different sequence K_N of trigonometric polynomials that have the same three properties. (See Exercise 4, Section 13H.)

2. The Weierstrass Polynomial Approximation Theorem proved in Section 7D can be derived easily from Theorem 13.11. In fact, suppose that g is continuous on an interval $[a, b]$. By rescaling and translating, we may assume that $-\pi < a < b < \pi$. Then there is a continuous periodic function f that agrees with g on $[a, b]$. It can be approximated uniformly within a given $\varepsilon/2$ by a trigonometric polynomial. The power series expansions of the exponentials e_n converge uniformly on $[a, b]$, so we may replace each exponential by a suitable Taylor polynomial and thus

replace the trigonometric polynomial by an ordinary polynomial that approximates it within $\varepsilon/2$ on the interval $[a, b]$. Thus we have approximated f within ε on this interval.

Exercises

1. Suppose that f is an integrable periodic function and suppose that $\widehat{f}(n) = 0$, all n. Prove that $f = 0$ a.e.
2. Obtain the results in Exercise 1 of Section 13D for integrable periodic f and g.

13F. L^2-Periodic Functions: The Riesz-Fischer Theorem

Fejér's Theorem shows one kind of "mean convergence" of the partial sums $S_N f(x)$: convergence of the arithmetic means. A second kind of mean convergence refers to averaging with respect to x by integration. The discussion in Section 13A already suggested that there is a close fit between Fourier series and the space $L^2(I)$. The next theorem spells that out in detail.

Theorem 13.12: The Riesz-Fischer Theorem. *Suppose that f is an L^2-periodic function. Then the partial sums of its Fourier series converge to f in the sense of $L^2(I)$:*

$$\lim_{N \to \infty} ||S_N f - f|| = 0. \tag{42}$$

Moreover, there is an equality (Parseval's Identity):

$$\sum_{n=-\infty}^{\infty} |\widehat{f}(n)|^2 = ||f||^2. \tag{43}$$

Conversely, suppose that $(a_n)_{-\infty}^{\infty}$ is a two-sided complex sequence that is square-summable, that is, $\sum |a_n|^2 < \infty$. Then there is a unique function f in $L^2(I)$ that has the a_n's as its Fourier coefficients.

Proof: If f is continuous, then Fejér's Theorem shows that $T_N f$ converges to f uniformly, so it converges in $L^2(I)$. It follows that the $S_N f$'s must also converge to f in $L^2(I)$: See Exercise 9 of Section 12F. For general f in $L^2(I)$, given $\varepsilon > 0$, choose a continuous periodic g such that $||f - g|| < \varepsilon$. Then we use (24) to estimate

$$||S_N f - f|| \leq ||S_N(f - g)|| + ||S_N g - g|| + ||g - f||$$
$$\leq ||f - g|| + ||S_N g - g|| + ||g - f|| \leq ||S_N g - g|| + 2\varepsilon,$$

which is $< 3\varepsilon$ for large N. This proves (42). The identity (23), in combination with (22), implies the identity

$$\|f\|^2 = \|S_N f\|^2 + \|f - S_N f\|^2 = \sum_{n=-N}^{N} |\widehat{f}(n)|^2 + \|f - S_N f\|^2. \quad (44)$$

The identity (43) is a consequence of (44) and (42),

Conversely, suppose that $(a_n)_{-\infty}^{\infty}$ is a square-summable sequence. Set

$$f_N(x) = \sum_{n=-N}^{N} a_n e^{inx}.$$

The orthonormality of the exponential functions e_n implies that, for $M < N$,

$$\|f_N - f_M\|^2 = \sum_{M < |n| \le N} |a_n|^2. \quad (45)$$

By the assumption of square summability, the right side converges to zero as M, N tend to ∞, so $(f_N)_0^{\infty}$ is a Cauchy sequence in $L^2(I)$. Let f be the limit. Orthonormality implies that $(f_N, e_n) = a_n$ for all $N \ge |n|$. This fact and the Cauchy-Schwarz inequality imply

$$\widehat{f}(n) = (f, e_n) = \lim_{N \to \infty} (f_N, e_n) = a_n. \quad \square \quad (46)$$

As noted above, the Fourier partial sums $S_N f(x)$ do not necessarily converge to f at a given point x, even if f is continuous. L. Carleson proved in 1966 that convergence occurs almost everywhere if f belongs to $L^2(I)$ (and, therefore, if f is continuous). The proof is quite difficult and technical.

The equality (43) is a special case of an equality for the inner product in the Hilbert space $L^2(I)$, also known as Parseval's Identity.

Theorem 13.13. *If f and g belong to $L^2(I)$, then*

$$(f, g) = \lim_{N \to \infty} \sum_{n=-N}^{N} \widehat{f}(n) \overline{\widehat{g}(n)}, \quad (47)$$

or, more succinctly, $(f, g) = \sum \widehat{f}(n) \overline{\widehat{g}(n)}$.

Proof: Again, the Cauchy-Schwarz inequality and the orthonormality of the e_n's imply that

$$(f, g) = \lim_{N \to \infty} (S_N f, S_N g) = \lim_{N \to \infty} \sum_{n=-N}^{N} \widehat{f}(n) \overline{\widehat{g}(n)}. \quad \square$$

13F. L^2-Periodic Functions: The Riesz-Fischer Theorem

Another way to look at the results in this section is in terms of standard models of various spaces. The standard model of an n-dimensional complex inner product space is the space $\mathbb{C}^n$ with standard inner product

$$(\mathbf{z}, \mathbf{w}) = \sum_{j=1}^{n} z_j \bar{w}_j.$$

Suppose that V is another n-dimensional complex inner product space and $\{\mathbf{e_1}, \ldots, \mathbf{e_n}\}$ is an orthonormal basis for V. Consider the linear transformation that takes a vector $\mathbf{v} \in V$ to the vector whose entries are its coefficients $a_j = (\mathbf{v}, \mathbf{e_j})$ with respect to the orthonormal basis. This linear transformation allows us to "identify" V with $\mathbb{C}^n$. (More formally, it is a unitary transformation: an isomorphism from V to $\mathbb{C}^n$ that preserves the inner products.)

Here the role of a standard model of an infinite-dimensional (separable) complex Hilbert space is taken by the space of two-sided complex sequences $\{a_n\}_{-\infty}^{\infty}$ that are square-summable:

$$\sum_{n=-\infty}^{\infty} |a_n|^2 < \infty.$$

The inner product between two such sequences is defined to be

$$\sum_{n=-\infty}^{\infty} a_n \bar{b}_n.$$

The Riesz-Fischer Theorem says that $L^2(I)$ may be identified with this sequence space by identifying a function f with its sequence of Fourier coefficients.

Exercises

1. Prove that for any f in $L^2(I)$ and any $N = 0, 1, 2, \ldots$, the partial sum $S_N f$ is the best approximation to f in the subspace spanned by the functions $\{e_n : |n| \leq N\}$.
2. Suppose that the sequence $(f_n)_{n=1}^{\infty}$ converges to f in the space $L^1(I)$. Prove that

$$\lim_{n \to \infty} \widehat{f}_n(k) = \widehat{f}(k), \quad \text{all } k \in \mathbb{Z}.$$

3. Suppose that f_n converges to f and g_n converges to g in $L^2(I)$. Prove that

$$\lim_{n \to \infty} (f_n, g) = (f, g), \quad \lim_{n \to \infty} (f_n, g_n) = (f, g).$$

4. Use Parseval's Identity (43) and the functions $f(x) = x$ and $g(x) = x^2$ to compute

   (a) $$\sum_{n=1}^{\infty} 1/n^2,$$

   (b) $$\sum_{n=1}^{\infty} 1/n^4.$$

13G. More Convergence

The Riesz-Fischer Theorem shows how to create a function in $L^2(I)$ with appropriate Fourier coefficients. The following theorem gives sufficient (but not necessary) conditions on the Fourier coefficients of a *continuous* periodic function.

Theorem 13.14. *Suppose that* $\{a_n\}_{-\infty}^{\infty}$ *is a complex sequence such that*

$$\sum_{n=-\infty}^{\infty} |a_n| < \infty.$$

Then the partial sums

$$f_N(x) = \sum_{-N}^{N} a_n e^{2\pi i n x}$$

converge uniformly to a continuous periodic function f whose Fourier coefficients are $\widehat{f}(n) = a_n$.

Proof: Since the e_n's have modulus 1, if $M < N$, then

$$|f_N(x) - f_M(x)| \leq \sum_{M < |n| \leq N} |a_n|,$$

and the right side tends to 0 as $M, N \to \infty$ by the assumption. Therefore, the f_N's converge uniformly and it follows that the limit f is continuous; it is periodic since the f_N's are. The conclusion about the Fourier coefficients follows as in the last part of Theorem 13.12, by taking limits. □

Lemma 13.15. *Suppose that* $(a_n)_{-\infty}^{+\infty}$ *is a complex sequence such that*

$$\sum_{n=-\infty}^{\infty} |n a_n|^2 < \infty.$$

Then

$$\sum_{n=-\infty}^{\infty} |a_n| < \infty.$$

Proof: We may assume that $a_n = 0$. By the discrete Cauchy-Schwarz inequality,

$$\left(\sum_{n=-\infty}^{\infty} |a_n|\right)^2 \leq \sum_{n=-\infty}^{\infty} |n a_n|^2 \cdot 2 \sum_{n=1}^{\infty} \frac{1}{n^2}. \quad \square$$

Theorem 13.16. *Suppose that f is continuous and periodic, and suppose that the derivative f' exists and is continuous. Then $S_N f$ converges uniformly to f.*

Proof: Use Exercise 4 of Section 13G and the preceding lemma. □

As a consequence of (the proof of) Fejér's Theorem, we can obtain a useful result on convergence of the Fourier series itself.

Theorem 13.17. *Suppose that f is an integrable periodic function. Suppose that f is continuous at the point x and suppose that the Fourier series of f converges at the point x, that is, suppose that $\lim_{N\to\infty} S_N f(x)$ exists. Then this limit is equal to $f(x)$.*

Sketch of proof: The argument, in the proof of Fejér's Theorem for convergence of $T_N f$ to f at the point x, goes through under these assumptions on f. The reason is that the translates f_a converge to f with respect to the L^1 norm as $a \to 0$ (Exercise 3 of Section 12C). If $S_N f(x)$ converges, then the averages $T_N f(x)$ converge to the same limit (Exercise 3 of Section 3D), so $\lim_{N\to\infty} S_N f(x) = f(x)$. □

We note here, without proof, a result concerning convergence when the function is not assumed to be continuous.

Theorem 13.18: Lebesgue's Theorem. *If f is an integrable periodic function, then $\lim_{N\to\infty} T_N f(x) = f(x)$ a.e.*

Corollary 13.19. *Suppose that f_1 and f_2 are integrable periodic functions. Then f_1 and f_2 have the same sequence of Fourier coefficients if and only if $f_1(x) = f_2(x)$ a.e.*

Proof: Let $f = f_1 - f_2$. If $f = 0$ a.e., then $\widehat{f}(n) = 0$, all n. Conversely, suppose that $\widehat{f}(n) = 0$, all n. One way to see that this implies that $f = 0$ a.e. is to note that $T_N f \equiv 0$ for every N and use Lebesgue's Theorem. For another way, see Exercise 1 of Section 13E. □

Exercises

1. Are any of the following sequences the Fourier coefficients of a continuous periodic function? A continuously differentiable periodic function? An L^2-periodic function?

   (a) $a_n = \dfrac{1}{1+n^2}$.

   (b) $b_n = \dfrac{(-1)^n}{1+|n|}$.

   (c) $c_n = \dfrac{(-1)^n}{\sqrt{1+|n|}}$.

   (d) $d_n = \dfrac{1}{(1+n^2)^{2/3}}$.

2. Let $f(x) = \sum_{n=0}^{\infty} \cos nx/(1+n^2)$.
   (a) Show that f is continuous and periodic.
   (b) Determine the Fourier coefficients of f.
   (c) Prove or disprove: f has continuous first and second derivatives.
3. Show that for each r such that $0 \leq r < 1$, there is a function f whose Fourier coefficients are $\widehat{f}(n) = 0$ for $n < 0$ and $\widehat{f}(n) = r^n$ for $n \geq 0$. Calculate f.
4. Show that for each $0 \leq r < 1$, there is a function f_r whose Fourier coefficients are $\widehat{f_r}(n) = r^{|n|}$. Show that

$$f_r(x) = \frac{1-r^2}{1+r^2 - 2r\cos 2\pi x}.$$

5. Suppose that f is continuous and periodic and that it has a continuous derivative.
   (a) Show that there is a constant C such that

$$|\widehat{f}(n)| \leq \frac{C}{|n|}, \quad n = \pm 1, \pm 2, \ldots.$$

   (b) Show that $\sum_{-\infty}^{+\infty} n^2 |\widehat{f}(n)|^2 < \infty$.
6. Suppose that f is an integrable periodic function whose Fourier coefficients satisfy $\sum_{-\infty}^{+\infty} n^2 |\widehat{f}(n)|^2 < \infty$.
   (a) Prove that there is a continuous function g such that $f = g$ a.e.
   (b) Prove that the derivatives $(S_N f)'$ converge in $L^2(I)$.
7. Suppose that f is continuous and periodic, and let $F(x) = \int_0^x f(x) dx$. Show that F is periodic if and only if $\widehat{f}(0) = 0$. If so, relate the Fourier coefficients of F to those of f.
8. Suppose that f is periodic and has continuous derivatives of order $\leq k$. Prove that there is a constant C such that

$$|\widehat{f}(n)| \leq \frac{C}{|n|^k}, \quad n = \pm 1, \pm 2, \ldots.$$

9. Suppose that f is an integrable periodic function, and suppose that for some constant C and some nonnegative integer k the Fourier coefficients satisfy

$$|\widehat{f}(n)| \leq \frac{C}{|n|^{k+2}}, \quad n = \pm 1, \pm 2, \ldots.$$

   Show that there is a function g such that g and its derivatives of order $\leq k$ are continuous, and such that $f = g$ a.e.
10. Suppose that f is periodic and both f and its derivative f' are bounded and are continuous except at isolated points; suppose that at such a point x_0 both the limit from the left and the limit from the right exist:

$$f(x_0-) = \lim_{x \to x_0, x > x_0} f(x); \quad f(x_0+) = \lim_{x \to x_0, x > x_0} f(x).$$

Show that the Fourier series at x_0 converges to $\frac{1}{2}[f(x_0-) + f(x_0+)]$.

13H*. Convolution

The expressions for the partial Fourier sum $S_N f$ and its sequence of averages $T_N f$ suggest an operation that assigns a periodic function to a pair of periodic functions f and g: the convolution product, or simply *convolution* $f * g$. Formally, it is defined by

$$[f * g](x) = \frac{1}{2\pi} \int_{-\pi}^{\pi} f(y) g(x-y) dy.$$

As in (7),

$$[f * g](x) = \frac{1}{2\pi} \int_{-\pi}^{\pi} f(x-y) g(y) dy = [g * f](x).$$

This makes sense if one of the two periodic functions is continuous (or piecewise continuous) and the other is integrable. In fact, it can be shown to be well defined for a.e. x if we only assume that both f and g are integrable periodic functions. In particular, we have, for any periodic f in $L^1(I)$,

$$S_N f = D_N * f = f * D_N, \qquad T_N f = F_N * f = f * F_N,$$

where D_N is the Dirichlet kernel and F_N is the Fejér kernel.

The Fejér kernel is one example of the important notion of an *approximate identity*. A sequence $(\varphi_n)_{n=1}^{\infty}$ of continuous periodic functions is said to be an approximate identity if it satisfies the three conditions that we identified for $(F_n)_{n=1}^{\infty}$:

(i) $\qquad\qquad\qquad\qquad\qquad \varphi_n \geq 0;$

(ii) $\qquad\qquad\qquad \frac{1}{2\pi} \int_{-\pi}^{\pi} \varphi_n(y) dy = 1, \qquad$ for all $n \in \mathbb{N}$;

(iii) $\displaystyle\lim_{n \to \infty} \left[\int_{-\pi}^{-\delta} \varphi_n(y) dy + \int_{\delta}^{\pi} \varphi_n(y) dy \right] = 0 \qquad$ for any fixed δ, $0 < \delta < \pi$.

As we noted in connection with the Weierstrass Approximation Theorem, the conclusion of Fejér's Theorem, and its proof, remain valid for any approximate identity: If f is a continuous periodic function, then $f * \varphi_n$ converges to f uniformly as $n \to \infty$.

Exercises

1. Suppose that f and g are continuous periodic functions.
   (a) Show that the Fourier coefficients of the convolution $f * g = h$ satisfy

   $$\widehat{h}(k) = \widehat{f}(k) \widehat{g}(k), \qquad \text{all } k \in \mathbb{Z}.$$

   (b) Interpret this when $g = D_N$, the Dirichlet kernel.

(c) Show that this remains true if it is only assumed that f is integrable and g is square integrable.

2. Suppose that $(g_n)_{n=1}^{\infty}$ is an approximate identity. Show that for each integer k,
$$|\widehat{g_n}(k)| \leq 1 \quad \text{and} \quad \lim_{n \to \infty} \widehat{g_n}(k) = 1.$$
Check this with $g_N = F_N$, the Fejér kernel.

3. Use the preceding two exercises and Parseval's Identity to show that if f is an L^2-periodic function and $(g_n)_{n=1}^{\infty}$ is an approximate identity, then
$$\lim_{n \to \infty} ||g_n * f - f|| = 0.$$

4. Show that constants c_n can be chosen so that the sequence of functions
$$g(x) = c_n [1 + \cos x]^n, \quad n = 0, 1, 2, \ldots.$$
is an approximate identity. Compute the c_n's. (This leads to the original proof by Weierstrass of his approximation theorem.)

14*

Applications of Fourier Series

In this chapter we analyze a famous phenomenon of Fourier expansions, touch on a few of the many uses of Fourier series, and introduce the related topics of finite Fourier expansions and the Fourier integral.

14A*. The Gibbs Phenomenon

Although the Fourier series of a piecewise continuous function like the *square-wave* function

$$f(x) = \begin{cases} -1, & -\pi \leq x < 0; \\ 1, & 0 \leq x < \pi \end{cases}$$

converges at each point of continuity of f, by Dirichlet's Theorem, it was noticed by Wilbrahim and others, including J. W. Gibbs, that the partial sums $S_N f$ overshoot for small positive x and undershoot for small negative x by an amount that tends rapidly to a nonzero constant as $N \to \infty$. See Figure 11.

We sketch here a precise analysis of this phenomenon and an evaluation of the constant. Direct calculation shows that the Fourier coefficients of the square-wave function f are

$$\widehat{f}(2n) = 0; \qquad \widehat{f}(2n+1) = \frac{2}{i(2n+1)\pi}.$$

Therefore,

$$S_{2n-1} f(x) = \frac{4}{\pi} \left[\sin x + \frac{\sin 3x}{3} + \cdots + \frac{\sin(2n-1)x}{2n-1} \right]$$

$$= \frac{4}{\pi} \int_0^x [\cos t + \cos 3t + \cdots + \cos(2n-1)t] \, dt$$

$$= \frac{4}{\pi} \int_0^x \operatorname{Re} \left[e^{it} + e^{i3t} + \cdots + e^{i(2n-1)t} \right] dt$$

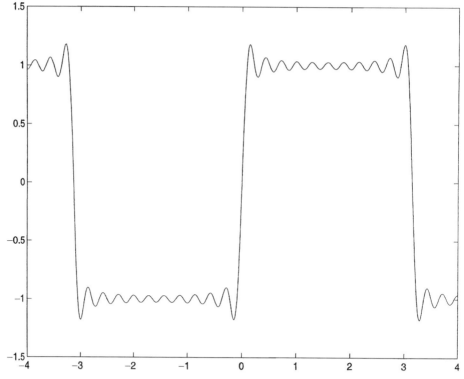

Figure 11. S_N of the square wave function for $N = 21$.

$$= \frac{4}{\pi}\int_0^x \operatorname{Re}\left[\frac{e^{i2nt}-1}{e^{it}-e^{-it}}\right]dt = \frac{4}{\pi}\int_0^x \operatorname{Re}\left[\frac{e^{-2nt}-1}{2i\sin t}\right]dt$$
$$= \frac{2}{\pi}\int_0^x \frac{\sin 2nt}{\sin t}\,dt. \tag{1}$$

We are interested in small positive x, so the last integral is approximately

$$\frac{2}{\pi}\int_0^x \frac{\sin 2nt}{t}\,dt = \frac{2}{\pi}\int_0^{2nx} \frac{\sin s}{s}\,ds.$$

The extrema of $\int_0^y \sin s\,ds/s$ for positive y occur at integer multiples of π, and since the humps of the function being integrated decrease in size, and their signs alternate, the maximum occurs at $y = \pi$. Therefore the maximum of (1) for $0 < x < 1/2$ and n large occurs near $x = \pi/2n$ and has a value approximately equal to

$$\frac{2}{\pi}\int_0^\pi \frac{\sin s}{s}\,ds = 1.179\ldots.$$

In other words, the overshoot, in the limit, is nearly 18 percent.

Exercises

1. Suppose that f is a piecewise continuous function as in Exercise 10 of Section 13G. Discuss the behavior of $S_N f$ near a point of discontinuity of f for large N. Is there an undershoot or overshoot to the right of the point? By how much, and why?
2. Do the arithmetic means $T_N f$ of the square-wave function (1) exhibit an overshoot? Why, or why not?

14B*. A Continuous, Nowhere Differentiable Function

Filling in the gaps of an idea of Riemann, Weierstrass distressed various nineteenth-century mathematicians by producing a continuous, periodic, nowhere differentiable function with a simple Fourier series representation. In fact, choose a in $(0, 1)$ and consider

$$f(x) = \sum_{0}^{\infty} a^n \cos(2^n x) \qquad 0 < a < 1. \tag{2}$$

This function is continuous, by Theorem 13.14.

Theorem 14.1. *The function f of (2) is not differentiable at any point if $a > 1/2$.*

(In fact, a modification of the argument below, using the square of the Fejér kernel, shows that f is also nowhere differentiable if $a = 1/2$. The result was proved by Weierstrass for a close to 1 and extended down to $a = 1/2$ by G. H. Hardy.)

Theorem 14.1 is a consequence of the following result.

Theorem 14.2. *Suppose that g is periodic and continuous and suppose that the Fourier coefficients $\widehat{g}(m)$ vanish unless $m = \pm 2^n$ for some integer $n \geq 0$, and $\widehat{g}(\pm 2^n) = a_n^{\pm}$. If g is differentiable at some point x_0, then there is a constant C such that*

$$|a_n^{\pm}| \leq C n 2^{-n}, \qquad n = 1, 2, 3, \ldots. \tag{3}$$

Proof: Translating g by x_0 does not change the moduli of the Fourier coefficients (Proposition 13.3), so we may assume that g is differentiable at 0. Subtracting a constant from g changes only $\widehat{g}(0)$, so we may assume that $g(0) = 0$. These two assumptions imply that, for some constant M_0,

$$|g(x)| \leq M_0 |x|, \qquad \text{all } x.$$

This in turn implies that, for some constant M,

$$h(x) = \frac{|g(x)|}{|\sin^2 \tfrac{1}{2} x|} \leq \frac{M}{|x|}, \qquad \text{all } x \in (-\pi, \pi), \quad x \neq 0. \tag{4}$$

Let us prove the inequality (3) for a_n^+. The assumptions on the Fourier coefficients imply that

$$(g, e_m) = 0, \quad \text{if } 2^{n-1} < m < 2^{n+1} \text{ and } m \neq 2^n; \quad (g, e_{2^n}) = a_n^+.$$

This implies that, for the Fejér kernel F_N,

$$a_n^+ = (g, e_{2^n} F_N), \quad N = 2^{n-1} - 1.$$

Using the form of the Fejér kernel (equation (37) of Chapter 13) and the inequalities (4) and $|\sin t| \leq |t|$, we get

$$|a_n^\pm| \leq \frac{1}{2\pi} \int_{-\pi}^{\pi} \frac{M \sin^2(\frac{N}{2}x)}{N|x|} dx$$

$$\leq \frac{M}{N\pi} \int_0^{\frac{1}{N}} 4N^2 |x| \, dx + \frac{M}{N\pi} \int_{\frac{1}{N}}^{\pi} \frac{dx}{|x|}$$

$$\leq M_1 \left[\frac{1}{N} + \frac{\log N}{N} \right].$$

Since $N = 2^{n-1}$, this gives the desired inequality. □

Note that the proof did not use the full force of differentiability; it only used boundedness, for some x_0, of the difference quotient

$$\frac{|g(x) - g(x_0)|}{|x - x_0|}.$$

Exercise

1. Let f be the Weierstrass function (2). Prove that if $a = 2^{-\alpha}$, where $0 < \alpha \leq 1$, then there is a constant M such that

$$|f(x) - f(y)| \leq M|x - y|^\alpha, \quad \text{all } x, y.$$

In particular, the difference quotients are bounded when $a = 1/2$, even though (by Hardy's theorem) f is nowhere differentiable.

Note that this exercise shows that the result in Exercise 7 of Section 13B is optimal.

14C*. The Isoperimetric Inequality

An unexpected application of Fourier series is the following.

Theorem 14.3. *Let γ be a smooth curve in the plane that bounds a region A and suppose that γ has length 2π. Then the area of A is π if γ is a circle and strictly less than π if γ is not a circle.*

14C*. The Isoperimetric Inequality

In other words, this confirms the intuitive feeling that, among curves of a given length, the circle encloses the largest total area.

The first step of the proof is to assume that the curve is oriented so that A lies to the left; then, by Green's Theorem,

$$\int_\gamma x\,dy - y\,dx = \int\int_A \left[\frac{\partial}{\partial x}x - \frac{\partial}{\partial y}(-y)\right]dx\,dy = 2\int\int_A dx\,dy = 2\,\text{area}(A). \quad (5)$$

Next, we may assume that the curve is parameterized by arc length, so that

$$\gamma(t) = (x(t), y(t)), \quad 0 \leq t \leq \pi; \quad [x'(t)]^2 + [y'(t)]^2 = 1. \quad (6)$$

Then x and y may be considered as periodic functions with Fourier coefficients

$$\widehat{x}(n) = a_n, \quad \widehat{y}(n) = b_n.$$

We interpret the line integrals in (5) as an inner products and use the expression of an inner product in terms of Fourier coefficients (equation (47) of Chapter 13) and the relation of the Fourier coefficients of a function to those of its derivative (equation (18) of Chapter 13) to obtain

$$\int_\gamma x\,dy - y\,dx = \int_0^{2\pi}[x(t)y'(t) - x'(t)y(t)]\,dt = 2\pi[(y', x) - (x', y)]$$

$$= 2\pi \sum_{n=-\infty}^\infty in[b_n\bar{a}_n - a_n\bar{b}_n]. \quad (7)$$

Using (5), (7), and the elementary inequality $|b\bar{a} - a\bar{b}| \leq 2|ab| \leq |a|^2 + |b|^2$, we obtain

$$2\,\text{area}(A) \leq \sum_{n=-\infty}^\infty 2\pi|n|(|a_n|^2 + |b_n|^2) \leq \sum_{n=-\infty}^\infty 2\pi n^2(|a_n|^2 + |b_n|^2),$$

with *strict* inequality unless $a_n = b_n = 0$ for all $|n| > 1$. Now we make use of the identity for the derivatives in (6) to see that

$$1 = (x', x') + (y', y') = \sum_{n=-\infty}^\infty n^2(|a_n|^2 + |b_n|^2). \quad (8)$$

Combining (7) and (8), we get

$$2\,\text{area}(A) \leq 2\pi,$$

with *strict* inequality unless $a_n = b_n = 0$ for all $|n| > 1$. Since the functions x, y are real, equality implies that x' and y' are linear combinations of $\cos t$ and $\sin t$. If we choose as our starting point $t = 0$, the point where x is maximal, it follows that $x'(0) = 0$ and $x''(0) < 0$ and thus $x'(t) = -\sin t$. Then, since $(x')^2 + (y')^2 \equiv 1$, it follows that $y'(t) = \pm\cos t$, and the curve is a circle.

14D*. Weyl's Equidistribution Theorem

Let $[x]$ denote the *integer part* of the real number x, that is, the unique integer m such that $m \leq x < m+1$. Let $\{x\}$ denote the *fractional part*: $\{x\} = x - [x] \in [0, 1)$. If x is rational, then the fractional parts of its integer multiples, $\{nx\}$, take only finitely many distinct values. Conversely, if x is not rational, then it is easy to see that the $\{nx\}$'s take infinitely many values; in fact, no two are the same. Kronecker showed that for irrational x these values are *dense* in the interval $[0, 1)$, that is, each open subinterval contains $\{nx\}$ for some integer n (and therefore for infinitely many choices of n). One can ask for more information: On average, over the long run, what fraction of the values lie in a given subinterval (a, b)? If we denote the number of elements in a finite set A by $\text{card}(A)$, the question is, does the limit

$$\lim_{N \to \infty} \frac{1}{2N+1} \text{card} \{n : |n| \leq N \text{ and } \{nx\} \text{ in } (a, b)\}, \tag{9}$$

exist and, if so, what is it? A two-sided sequence in the interval $[0, 1)$ is said to be *equidistributed* if the corresponding limit exists and is equal to $b - a$ for every subinterval (a, b). In other words, for an equidistributed sequence, in the long run the likelihood of its lying in a given subinterval is equal to its length. Weyl showed that for any irrational x the sequence with $x_n = \{nx\}$ is equidistributed. Here is a sketch of the proof.

First, we rewrite the quantity in (9), using the indicator function 1_A of the interval $A = (a, b)$; then this quantity is

$$\frac{1}{2N+1} \sum_{n=-N}^{N} 1_A(x_n).$$

Note the resemblance to a Riemann sum. The equidistribution property can be written

$$\lim_{N \to \infty} \frac{1}{2N+1} \sum_{n=-N}^{N} 1_A(x_n) = \int_0^1 1_A(y) dy. \tag{10}$$

If (10) is true for each subinterval A of $[0, 1)$, then it extends to linear combinations, that is, to step functions f:

$$\lim_{N \to \infty} \frac{1}{2N+1} \sum_{n=-N}^{N} f(x_n) = \int_0^1 f(y) dy. \tag{11}$$

If (11) is true for all step functions f, then an approximation argument shows that it is also true for all continuous functions, and in particular for all trigonometric polynomials scaled to have period 1.

Weyl's idea was to reverse this line of reasoning: If (11) is true for all such trigonometric polynomials, then the Weierstrass Approximation Theorem shows that it is true for all continuous functions, and a further approximation argument extends the validity to indicator functions of intervals. Thus one obtains the desired identity (10).

By linearity, (11) holds for trigonometric polynomials if it holds for the scaled exponentials $e_n(2\pi x)$. For $f = e_0$, both sides of (11) equal 1. For $f(x) = e_n(2\pi x)$, $n \neq 0$, the right side of (11) is 0. Thus the task is simply to show that in this case the left side is also 0. But

$$e_n(2\pi x_n) = e^{i2\pi x_n} = \exp(i2\pi \{nx\}) = \exp(i2\pi nx) = \alpha^n, \qquad \alpha = e^{i2\pi x}.$$

Since x is assumed irrational, it follows that $\alpha^n \neq 1$ for $n \neq 0$. Therefore,

$$\frac{1}{2N+1} \sum_{n=-N}^{N} e_n(2\pi x_n) = \frac{1}{2N+1} \frac{\alpha^{N+1} - \alpha^{-N}}{\alpha - 1}$$

and powers of α have modulus 1, so the limit is 0.

An analogous result with an analogous proof holds for several reals: If x_1, $x_2, \ldots, x_r$ are irrational real numbers that are linearly independent over the rationals, then the fractional parts of integral linear combinations

$$\{n_1 x_1 + n_2 x_2 + \cdots + n_r x_r\}, \qquad n_1, n_2, \ldots, n_r \in \mathbb{Z} \tag{12}$$

are dense in [0, 1). One consequence is that irrational flows on a torus are "ergodic."

Exercises

1. Prove that if (11) is true for each continuous f, then (10) is true for every subinterval A of [0, 1).
2. Formulate and prove the analogous equidistribution result for the fractional parts (12).

14E*. Strings

Historically, a direct antecedent to the idea of Fourier analysis was the analysis, by D'Alembert and Euler in the mid-eighteenth century, of the *vibrating string problem*. A string, say of length π, is stretched between two fixed points and is set moving by plucking, striking, or bowing. The problem is to describe mathematically the subsequent motion. If the motion is assumed to be in the vertical plane through the fixed ends that are at the same height, then the state of the string at time t is given by the function $u(x, t)$ that is the vertical displacement of the string above the rest position at distance x from the left endpoint. In particular, the fixed endpoints imply the *boundary conditions*

$$u(0, t) = u(\pi, t) = 0, \quad \text{all } t. \tag{13}$$

The equation of motion of the string is given by Newton's general law: Mass times acceleration equals force. For a taut string one may essentially ignore gravity and take the force to result from the tension, which for small displacements is proportional, at a point on the string, to the amount the string curves at that point – in other words, to the second partial derivative with respect to x. The acceleration is the second time derivative. Thus, for small displacements and ignoring friction, the equation of motion has the form

$$\frac{\partial^2 u}{\partial t^2} = c^2 \frac{\partial^2 u}{\partial x^2}, \tag{14}$$

where c is a constant that depends on density and tension. If the string is set in motion at time $t = 0$, then (2) should hold for $t > 0$ and $0 < x < 1$. To complete the mathematical description of the problem, one should give the *initial conditions*, which in this case are the initial configuration and the initial state of motion

$$u(x, 0) = f(x), \quad \frac{\partial u}{\partial t}(x, 0) = g(x), \tag{15}$$

where f and g are known functions; for example, $g = 0$ means the string was held in position f and then released, while $f = 0$ means that a string at rest was given an initial impetus g.

Our strategy is to *assume that there is a solution* and find a good way to analyze it. Suppose that we were trying to solve an ordinary differential equation,

$$\frac{d^2 u}{dt^2}(t) = c^2 A u(t), \tag{16}$$

where u takes values in a vector space V and A is a linear transformation, $A : V \to V$. If the matrix of A with respect to some basis is *diagonal*, then with respect to that basis equation (14) becomes a collection of simple scalar equations. The matrix of A with respect to a basis is diagonal if and only if each of the basis elements is an eigenvector of A: a nonzero vector v such that $Av = \lambda v$ for some scalar λ.

We can consider (14) to be in the form (16) if we take V to be a space of functions on the interval $(0, \pi)$ and take A to be the operator d^2/dx^2, which is indeed linear. Thus we would like to find a basis for our function space consisting of the *eigenfunctions* of d^2/dx^2: nonzero solutions v of

$$\frac{d^2 v}{dx^2}(x) = \lambda^2 v(x), \quad x \in (0, 1). \tag{17}$$

Because of (13) we want also to impose the conditions

$$v(0) = v(\pi) = 0. \tag{18}$$

The solutions of (17) for $\lambda = 0$ have the form $ax + b$, but (18) implies $a = b = 0$. Otherwise, the solutions necessarily have the form

$$v(x) = ae^{\lambda x} + be^{-\lambda x}$$

(see the next chapter) and (18) implies that $a + b = 0$ and $ae^{\lambda} + be^{-\lambda} = 0$. Together these conditions imply that $\lambda = in\pi$ and v is a multiple of $\sin nx$. Thus we are led to try to expand u in the form

$$u(x, t) = \sum_{n=1}^{\infty} a_n(t) \sin nx. \tag{19}$$

We can deduce that such an expansion exists, under the assumption that u is square integrable with respect to x for each fixed t, by the following argument. Any function in $L^2((0, \pi))$ can be extended so as to be an odd function on the interval $(-\pi, \pi)$ and then can be extended to be be periodic with period 2π. Such functions can be expanded in Fourier series using the exponentials e_n, or, equivalently, the trigonometric functions $\cos nx$, $\sin nx$. Since we have taken our functions to be odd, only the sine functions play a role.

We differentiate both sides of (19) twice with respect to t (we simply *assume* that the right side can be differentiated term by term) and equate coefficients to see that the differential equation (14) leads to the equations

$$a_n''(t) = -n^2 c^2 a_n(t), \tag{20}$$

whose solutions can be written as

$$a_n(t) = b_n \cos nct + c_n \sin nct. \tag{21}$$

Putting (19) and (21) together, we find (formally) the most general form of the solution to (13), (14). The conditions (15) determine the constants b_n and c_n. In fact, extend f and g to be odd; then the b_n's are the coefficients of the sine expansion of f and the c_n's are multiples of the coefficients of the sine expansion of g.

Special solutions of the vibrating string problem can be obtained by taking a single term,

$$u_n(x, t) = \cos nct \sin nx. \tag{22}$$

These solutions are called *standing-wave* solutions; u_n moves back and forth between extreme positions as t varies; it has period $2\pi/nc$ and frequency (number of complete oscillations per unit time) equal to $nc/2\pi$. Thus the *characteristic frequencies* are the positive integer multiples of the lowest, or fundamental, frequency.

A little thought shows that the part of the solution associated to (21) can be written in a way that shows a closer connection to (22). In fact, take $(r_n, nc\theta_n)$ to be polar coordinates for (b_n, c_n), so that (21) becomes

$$r_n[\cos nct \cos nc\theta_n + \sin nct \sin nc\theta_n] = r_n \cos(nct - nc\theta_n).$$

Combining this with (22) and (16), we get

$$u(x, t) = \sum_{n=1}^{\infty} r_n u_n(x, t - \theta_n),$$

a sum of standing waves with time shifts.

So far we have shown that the solution of the problem (13), (14), (15) can be expressed as a sum of standing-wave solutions. Another look at the formulas shows that it can also be expressed as a sum of two *traveling-wave* solutions. In fact, the trigonometric identities

$$2 \cos nct \sin nx = \sin n(x - ct) + \sin n(x + ct),$$
$$2 \sin nct \sin nx = \cos n(x - ct) - \cos n(x + ct)$$

(which are easily obtained by using the complex exponential expressions) imply that the sum (19), (21) can be reorganized into the form

$$u(x, t) = f_1(x - ct) + f_2(x + ct) \tag{23}$$

for suitable functions f_1 and f_2, periodic with period 2π.

This last expression could have been arrived at much more directly and leads to a different approach to the whole problem (13)–(15). In fact, we can factor the differential operator that appears in (14):

$$\frac{\partial^2}{\partial t^2} - c^2 \frac{\partial^2}{\partial x^2} = \left(\frac{\partial}{\partial t} - c\frac{\partial}{\partial x}\right)\left(\frac{\partial}{\partial t} + c\frac{\partial}{\partial x}\right) = \left(\frac{\partial}{\partial t} + c\frac{\partial}{\partial x}\right)\left(\frac{\partial}{\partial t} - c\frac{\partial}{\partial x}\right).$$

This factorization leads to the decomposition (23) of solutions of the differential equations (14). The functions f_1 and f_2 should then be chosen so that the boundary conditions (13) and initial conditions (15) are satisfied; see Exercise 2.

In light of the relative simplicity of this second method of attack on the problem (13)–(15), the reader may wonder why we bothered with the first method. One answer is that the first method, and variants of it, apply to a much wider class of problems than does the second method. A second answer is that the standing-wave solutions (22) and the characteristic frequencies $nc/2\pi$ are important features of the problem. In fact, the ear – or any physical detector – can detect frequencies only over a finite range, which means that only certain partial sums in (19) can actually be detected.

Exercises

1. Given a string, we can vary either the tension, which changes the constant c, or the length. Find the standing-wave solutions for a string of length L and discuss the effect of lowering the tension or lengthening the string on the sequence of characteristic frequencies.
2. Relate the functions f_1 and f_2 in (23) to the functions f and g in (15), after extending f and g to be odd with period 2π.
3. The *heat equation* or *diffusion equation* in one space dimension is

$$\frac{\partial u}{\partial t}(x,t) = \kappa \frac{\partial^2 u}{\partial x^2}(x,t), \qquad \kappa \text{ a positive constant.}$$

(This describes the temperature of a wire in an insulating medium. The length unit has been chosen so that the wire has length 2π and κ is the coefficient of thermal conductivity.) If we assume that the ends of the wire are joined, that is, that it forms a closed curve, then it is appropriate to think of u as being a periodic function of x.

Discuss the solution of this equation for u periodic in x and $t > 0$, given the value of $u(x, 0)$:
(a) Write the solution as a convolution.
(b) Find the limit of u as $t \to \infty$.
4. Discuss the temperature of a wire of length 2π in an insulating medium if the initial temperature distribution is known and the ends of the wire are kept at a fixed temperature; here we have dropped the assumption that the wire forms a closed curve.

14F*. Woodwinds

Guitars, zithers, and the like are not the only instruments that can be analyzed using Fourier series. For woodwinds, the results have some similarities and some differences. We begin with the flute. The sound is produced by the vibrations of a cylindrical column of air. Small vibrations turn out to be governed by the same equation as for the vibrating string, (14); here we take $u(x, t)$ to be the difference between the *pressure* at time t and at distance x along the column and the ambient pressure in the room. (Pressure is essentially constant across the cross section of the column). The column is set into motion by blowing across an open hole, and the length of the vibrating column is essentially the distance to the next open hole, or to the open end of the instrument if no other hole is open. At the open holes, or the end, the pressure is the same as the ambient pressure, so boundary conditions (13) are appropriate if the length is taken to be π. Then the standing-wave solutions and the characteristic frequencies are the positive integer multiples of the fundamental frequency. One effect of this is that if the flute is blown to produce a fundamental tone C, and then is blown rather harder with the same fingering, the note produced is C', an octave higher (twice the frequency).

Now consider a clarinet. Playing the same game, one goes from a fundamental tone of C to G', the fifth above the octave C' (triple the frequency). This suggests that there is something wrong with equation (14) for the clarinet – or what? In fact, equation (14) is still correct (approximately, as before), but the boundary conditions (13) are not. The column of air in the clarinet is set in motion using the reed at one end, and there is no reason for the pressure there to be the same as that of the ambient air. What does happen at a closed end is that the pressure cannot drop off or increase sharply as a function of x. If the closed end is at $x = 0$, then the appropriate boundary conditions are

$$\frac{\partial u}{\partial x}(0, t) = 0, \quad u(1, t) = 0. \tag{24}$$

We leave it as an exercise to show that the standing-wave solutions for the problem (14), (24) have the form

$$\cos\left(\left[n + \tfrac{1}{2}\right]ct\right) \cos\left(\left[n + \tfrac{1}{2}\right]x\right), \tag{25}$$

and to deduce that the characteristic frequencies are precisely the *odd* positive integer multiples of the fundamental. In particular, above a fundamental C we would expect to hear the note with $3 = \frac{3}{2} \cdot 2$ the frequency, that is, the fifth above the octave C'.

Finally, consider an oboe, and play the same game. Like a clarinet, the oboe has one closed and one open end and therefore would seem to correspond to the same boundary conditions (24). Nevertheless, the tone heard above a fundamental C is the octave C'. This must mean that this time there is something wrong with equation (14), and indeed there is, though not because of some occult properties of the air in an oboe (or of the oboist). In fact, an oboe is not cylindrical; it is an elongated cone, tapering nearly to a point. Since the cross section is not constant, we would expect the coefficients of the equation to show some dependence on the variable x. (In deriving (14) for a string we tacitly assumed that the string was homogeneous; if it has different properties along its length, then (14) should not be accurate.) The correct equation can be shown to have the form

$$\frac{\partial^2 u}{\partial t^2}(x, t) = c^2 \left[\frac{\partial^2 u}{\partial x^2}(x, t) + \frac{2}{x} \frac{\partial u}{\partial x}(x, t) \right].$$

This equation can be rewritten as

$$\frac{\partial^2}{\partial t^2}[xu(x,t)] = c^2 \frac{\partial^2}{\partial x^2}[xu(x,t)]. \qquad (26)$$

Therefore the function $w(x,t) = xu(x,t)$ satisfies the wave equation (14). Moreover, w satisfies the boundary conditions (13), so the standing-wave solutions and the characteristic frequencies (for this simplified model) are the same (up to the determination of the constant c) as for the string or the flute.

Percussion can be more complicated.

Exercises

1. Find the eigenfunctions of the operator d^2/dx^2 subject to the boundary conditions (24) and follow the method of Section 14E to show that the general solution of (14) with boundary conditions (1) is a sum of standing-wave solutions (25), translated in time. In particular, one needs to show that any function in $L^2((0,\pi))$ has an expansion in the appropriate cosine functions of x.
2. Verify that the equation that precedes (26) is equivalent to (26).

14G*. Signals and the Fast Fourier Transform

In practical applications a function, such as a signal, is given by a finite amount of discrete data: N numbers, where N may be quite large but is definitely finite. Thus let us consider functions on a finite set, taken, for convenience, to be complex-valued:

$$f : S \to \mathbb{C}, \qquad S = S_N = \{0, 1, 2, \ldots, N-1\}. \qquad (27)$$

Equivalently, f is an N-tuple of complex numbers, an element of $\mathbb{C}^N$. However, it is more convenient to think of the representation (27) and to write the variable x for an element of S. We denote the set of such functions by $\mathcal{F} = \mathcal{F}_N$. Then $\mathcal{F}$ is taken to be a finite-dimensional Hilbert space with the inner product and norm

$$(f, g) = \frac{1}{N} \sum_{x \in S} f(x)\overline{g(x)}; \qquad \|f\|^2 = \frac{1}{N} \sum_{x \in S} |f(x)|^2. \qquad (28)$$

We may also extend our functions to be defined on the integers and periodic of period N:

$$f : \mathbb{Z} \to \mathbb{C}; \qquad f(x+N) = f(x), \quad x \in \mathbb{Z}. \qquad (29)$$

We leave it as an exercise to show that the exponential functions

$$e_n(x) = e^{2n\pi i x/N}, \qquad n = 0, 1, 2, \ldots, N-1, \quad x \in \mathbb{Z} \qquad (30)$$

satisfy (29) and are an orthonormal basis for $\mathcal{F}$. Consequently, any function f in $\mathcal{F}$ has the finite Fourier expansion

$$f(x) = \sum_{x \in S} \widehat{f}(n) e_n(x), \qquad \widehat{f}(n) = (f, e_n). \tag{31}$$

This decomposition can be useful for applications, but one needs to be concerned with the number of computations necessary to compute it when N is large (e.g., 2^{10} or 2^{20}). Now

$$\widehat{f}(n) = \frac{1}{N} \sum_{x \in S} f(x) e_n(-x), \tag{32}$$

so it appears that N multiplications and N additions are necessary to compute each of the N coefficients – a total of $2N^2$ operations. However, the process can be reorganized so as to involve on the order of $N \log N$ operations, that is, on the order of $10 \cdot 2^{10}$ rather than 2^{20} if $N = 2^{10}$, a factor of 1028; the saving grows very rapidly with N. This is known as the *fast Fourier transform* (FFT). It was essentially known to Gauss in the early nineteenth century and was rediscovered in the twentieth century. Suppose that $N = 2M$ is even. Given f in $\mathcal{F}_N$, define the functions

$$f_+(x) = \frac{f(x) + f(x+M)}{2}; \qquad f_-(x) = \frac{f(x) - f(x+M)}{2} \cdot e_{-1}(x). \tag{33}$$

Both these functions have period $M = N/2$. In the case of f_-, this follows from the fact that $e_{-1}(x + M) = -e_{-1}(x)$. Now the functions e_{2n} are the orthonormal exponentials for $\mathcal{F}_M$. It follows that the Fourier coefficients of f_+ and f_- as elements of $\mathcal{F}_M$ are related to the Fourier coefficients of f by

$$\widehat{f}_+(k) = \widehat{f}(2k), \qquad \widehat{f}_-(k) = \widehat{f}(2k+1). \tag{34}$$

Thus the problem has been reduced to that of computing the Fourier coefficients of each of *two* functions in $\mathcal{F}_M$. If $N = 2^m$, the process can be iterated down to the point where we have N functions in $\mathcal{F}_1$, that is, N *constant* functions, whose values are exactly the Fourier coefficients of f.

In (34) we only need the values of $f_\pm$ at $0, 1, \ldots, M-1$, so there are $2M = N$ additions and (counting multiplication by $\frac{1}{2}$) $M + 2$ multiplications, one of which is trivial. Therefore, in at most 2N operations one reduces to two functions in $\mathcal{F}_M$, $M = n/2$. Inductively this means at most $2m \cdot 2^m$ operations when $N = 2^m$.

Exercises

1. Verify that the functions (30) are indeed an orthonormal basis for $\mathcal{F}$.
2. Verify that f_+ and f_- of (33) have period M.
3. Verify (34).

14H*. The Fourier Integral

Fourier series can be adapted to functions periodic of any period or, equivalently, to functions belonging to the Hilbert space $L^2(I)$ for any interval I. To see how to deal with functions on the whole line, the space $L^2(\mathbb{R})$, we pass to the limit from longer and longer intervals. Specifically, given $L > 0$, let $I_L = [-\frac{1}{2}, \frac{1}{2}L/]$ be the interval of length L centered at the origin. We take the inner product in $L^2(I_L)$ to be the (unnormalized) integral

$$(f, g)_L = \int_{L/2}^{L/2} f(x)\overline{g(x)}\, dx.$$

The rescaled complex exponentials

$$e_{nL}(x) = \frac{e^{i2\pi nx/L}}{\sqrt{L}}, \quad n \in \mathbb{Z} \tag{35}$$

are an orthonormal basis for $L^2(I_L)$.

Suppose that f belongs to the intersection $L^2(\mathbb{R}) \cap L^1(\mathbb{R})$. Let f_L denote its restriction to the interval I_L; thus

$$f_L \in L^2(I_L) \cap L^1(I_L). \tag{36}$$

Correspondingly, we have the Fourier expansion

$$f_L = \sum_{n=-\infty}^{\infty} \widehat{f_L}(n) e_{nL}; \tag{37}$$

$$\widehat{f_L}(n) = \frac{1}{\sqrt{L}} \int_{-\frac{1}{2}L}^{+\frac{1}{2}L} f(x) e^{-i2\pi nx/L}\, dx. \tag{38}$$

Let us associate to the sequence of Fourier coefficients of f_L a step function g_L, constant on intervals of length $2\pi/L$:

$$g_L(\xi) = \sqrt{L}\, \widehat{f_L}(n) = \int_{-\frac{1}{2}L}^{+\frac{1}{2}L} f(x) e^{-i2\pi nx/L}\, dx, \quad \text{for } \xi \in \left[\frac{2\pi n}{L}, \frac{2\pi(n+1)}{L}\right). \tag{39}$$

Then g_L is square integrable and

$$\frac{1}{2\pi} \int_{\mathbb{R}} |g_L(\xi)|^2 d\xi = \sum_{-\infty}^{\infty} |\widehat{f_L}(n)|^2 = \int_{-\frac{1}{2}L}^{+\frac{1}{2}L} |f(x)|^2 dx. \tag{40}$$

Because of our assumption that f is integrable on R, we may apply the Dominated Convergence Theorem to deduce from (39) that

$$\lim_{L \to \infty} g_L(\xi) = \widehat{f}(\xi), \quad \text{where } \widehat{f}(\xi) = \int_{\mathbb{R}} f(x) e^{-ix\xi} dx. \tag{41}$$

The function $\widehat{f}$ is called the *Fourier transform* of f.

For the moment let us assume more about the function f, say that it is continuously differentiable. Then at every interior point of the interval I_L the Fourier series of f_L converges and we can rewrite the resulting sum as an integral involving g_L:

$$\begin{aligned}
f(x) &= \sum_{n=-\infty}^{\infty} \frac{1}{\sqrt{L}} \widehat{f_L}(n) e_{nL}(x) \\
&= \sum_{-\infty}^{\infty} \frac{1}{L} g_L\left(\frac{2\pi n}{L}\right) e_{nL}(x) \\
&= \frac{1}{2\pi} \int_{\mathbb{R}} g_L(\xi) e_{Lx}(\xi) d\xi, \qquad |x| < \frac{1}{2} L,
\end{aligned} \qquad (42)$$

where e_{Lx} is the step function

$$e_{Lx}(\xi) = e^{i 2\pi n x / L}, \qquad \xi \in \left[\frac{2\pi n}{L}, \frac{2\pi (n+1)}{L}\right).$$

Taking the limit as $L \to \infty$ in (42) we have

$$f(x) = \frac{1}{2\pi} \int_{\mathbb{R}} \widehat{f}(\xi) e^{ix\xi} d\xi. \qquad (43)$$

The pair of formulas (41) and (43) show a certain duality between the function f and its Fourier transform $\widehat{f}$. Taking limits in (42) gives the relation

$$\int_{\mathbb{R}} |f(x)|^2 dx = \frac{1}{2\pi} \int_{\mathbb{R}} |\widehat{f}(\xi)|^2 d\xi. \qquad (44)$$

In fact, one can deduce a more general result about inner products in $L^2(\mathbb{R})$:

$$(f, g) = \frac{1}{2\pi} (\widehat{f}, \widehat{g}). \qquad (45)$$

So far we have obtained these formulas somewhat formally, and under the assumption that f is integrable and continuously differentiable, as well as square integrable. The precise version for square-integrable f is the following.

Theorem 14.4: Plancherel's Theorem. *Suppose that f belongs to L^2 on the line. Then, as L tends to $+\infty$, the functions*

$$\widehat{f_L}(\xi) = \int_{-\frac{1}{2}L}^{+\frac{1}{2}L} f(x) e^{-ix\xi} dx$$

converge in $L^2(\mathbb{R})$ to a function $\widehat{f}$. Conversely, the functions

$$f_L(x) = \frac{1}{2\pi} \int_{-\frac{1}{2}L}^{+\frac{1}{2}L} \widehat{f}(\xi) e^{ix\xi} \, d\xi$$

converge to f in $L^2(\mathbb{R})$. The identity (45) for inner products is valid for any functions f and g in $L^2(\mathbb{R})$.

We omit the details of the proof.

Suitably interpreted, or for suitable functions, we may write the preceding result in the form

$$\widehat{f}(\xi) = \int_\mathbb{R} e^{-ix\xi} f(x) \, dx; \quad f(x) = \frac{1}{2\pi} \int_\mathbb{R} e^{ix\xi} \widehat{f}(\xi) \, d\xi. \quad (46)$$

Theorem 14.5. *Suppose that f is a complex-valued function on the line that has continuous first and second derivatives, and suppose that f, f', and f'' are integrable. Then $\widehat{f}$ is integrable and equation (46) is valid at every point $x \in \mathbb{R}$.*

Proof: Clearly

$$|\widehat{f}(\xi)| \leq \int_\mathbb{R} |e^{-ix\xi} f(x)| \, dx = \int_\mathbb{R} |f(x)| \, dx. \quad (47)$$

It follows from Exercise 2, and the same calculation, that

$$|\xi^2 \widehat{f}(\xi)| \leq \int_\mathbb{R} |f''(x)| \, dx. \quad (48)$$

Therefore, $|\widehat{f}(\xi)| \leq C/(1+\xi^2)$ and it follows that $\widehat{f}$ is integrable. Now we introduce a "convergence factor" into the right side of (46), so that when the result is written as an iterated integral, the order of integration may be legitimately interchanged:

$$\frac{1}{2\pi} \int_\mathbb{R} e^{ix\xi} \widehat{f}(\xi) \, d\xi = \lim_{\varepsilon \to 0+} \frac{1}{2\pi} \int_\mathbb{R} e^{-(\varepsilon\xi)^2/2} e^{ix\xi} \widehat{f}(\xi) \, d\xi$$

$$= \lim_{\varepsilon \to 0+} \frac{1}{2\pi} \int_\mathbb{R} e^{-(\varepsilon\xi)^2/2} e^{ix\xi} \left[\int_\mathbb{R} e^{-iy\xi} f(y) \, dy \right] d\xi.$$

For fixed $\varepsilon > 0$ we can interchange the order of integration in the last expression and obtain, after some rewriting,

$$\frac{1}{2\pi} \int_\mathbb{R} e^{ix\xi} \widehat{f}(\xi) \, d\xi = \lim_{\varepsilon \to 0+} \int_\mathbb{R} G_\varepsilon(x-y) f(y) \, dy,$$

where

$$G_\varepsilon(x) = \frac{1}{2\pi} \int_{\mathbb{R}} e^{ix\xi} e^{-(\varepsilon\xi)^2/2} \, d\xi = \frac{1}{2\pi\varepsilon} \int_{\mathbb{R}} e^{i(x/\varepsilon)\xi} e^{-\xi^2/2} \, d\xi = \frac{1}{2\pi} \widehat{g}(-x/\varepsilon).$$

Here g is the Gaussian function $g(y) = e^{-y^2/2}$. According to Exercise 6 below, $G(x) = \frac{1}{2\pi} e^{-x^2/2}$. Thus

$$\frac{1}{2\pi} \int_{\mathbb{R}} e^{ix\xi} \widehat{f}(\xi) \, d\xi = \lim_{\varepsilon \to 0+} \int_{\mathbb{R}} G_\varepsilon(x-y) f(y) \, dy$$
$$= \lim_{\varepsilon \to 0} \int_{\mathbb{R}} \varepsilon^{-1} G(\varepsilon^{-1}(x-y)) f(y) \, dy.$$

From this point, the proof is exactly like the proof of Fejér's Theorem, since

$$G_\varepsilon \geq 0; \quad \int_{\mathbb{R}} G_\varepsilon(x) \, dx = 1; \quad \lim_{\varepsilon \to 0+} \int_{|x|>\delta} G_\varepsilon(x) \, dx = 0. \quad \square$$

Exercises

These exercises introduce some properties of the Fourier transform and some applications. One may assume that the functions in question satisfy enough hypotheses so that the formal manipulations are justified.

1. Continuity of the Fourier transform: If f is integrable, then $\widehat{f}$ is continuous.
2. The Fourier transform of the derivative: If $g = df/dx$, then $\widehat{g}(\xi) = i\xi \widehat{f}(\xi)$.
3. The Fourier transform under multiplication by x: If $h(x) = xf(x)$, then $d\widehat{f}(\xi)/d\xi = -i\widehat{h}(\xi)$.
4. The Fourier transform under translation: Find $\widehat{g}(\xi)$ if $g(x) = f(x-a)$, a fixed.
5. The Fourier transform under scaling: Find $\widehat{h}(\xi)$ if $h(x) = f(\rho x)$ for some positive ρ.
6. The Fourier transform of a Gaussian function: Let $f(x) = \frac{1}{2\pi} e^{-x^2/2}$. Show that $\widehat{f}(\xi) = e^{-\xi^2/2}$.
7. *The heat equation on the line*: To solve the problem

$$\frac{\partial u}{\partial t}(x,t) = \frac{\partial^2 u}{\partial x^2}(x,t), \quad x \in \mathbb{R}, \quad t > 0,$$
$$u(x,0) = f(x),$$

take the Fourier transform of u with respect to x and derive an equation for the transform $\widehat{u}(\xi,t)$, which implies that

$$\widehat{u}(\xi,t) = e^{-t\xi^2} \widehat{f}(\xi).$$

Show that the solution is given by

$$u(x,t) = \int_{\mathbb{R}} G(x-y,t) f(y) dy, \qquad G(x,t) = \frac{1}{\sqrt{4\pi t}} \cdot e^{-x^2/4t}.$$

8. *The Poisson summation formula*: Suppose that f is integrable on $\mathbb{R}$. Prove that

$$\sum_{m=-\infty}^{\infty} f(x - 2m\pi) = \frac{1}{2\pi} \sum_{n=-\infty}^{\infty} \widehat{f}(n) e^{inx},$$

under the assumption that f and its first derivative are continuous and converge rapidly enough to 0 as $|x| \to \infty$.

9. *Band-limited signals*: If we interpret x as the time variable, we may think of a function $f(x)$ as the amplitude of a signal. If $f(x) = \sin x\xi$, then this is a pure sine wave with frequency ξ. For a more general (but, say, integrable and square integrable) function f, the second half of equation (46) may be viewed as showing how f can be decomposed as a "continuous linear combination" of waves with frequencies ξ, the coefficients being $\widehat{f}(\xi)$. If the signal comes over a channel with limited bandwidth (meaning that there is a limit to the frequencies that can be transmitted – or detected), then the signal is completely determined by its values at a discrete set of points whose spacing depends on the bandwidth. Prove a precise form of this result: Suppose that $\widehat{f}(\xi) = 0$ for $|\xi| > M/2$. Show that f is continuous (assume 15.12 and use Dominated Convergence). Show that f is determined by its values at the points $2n\pi/M, n \in \mathbb{Z}$.

14I*. Position, Momentum, and the Uncertainty Principle

Suppose that f is a function in $L^2(\mathbb{R})$ with norm 1. Then $|f(x)|^2$, which has integral 1, can be thought of as a probability density, so that the probability that a point x lies in an interval I of $\mathbb{R}$ is

$$\text{Prob}\,\{x \in I\} = \int_I |f(x)|^2 dx. \tag{49}$$

Then the expected value of the position of the point is the mean

$$E = \int_{\mathbb{R}} x|f(x)|^2 dx. \tag{50}$$

The variance (square of the standard deviation) is a measure of how spread out the probability distribution is; it is

$$V = \int_{\mathbb{R}} (x-E)^2 |f(x)|^2 dx. \tag{51}$$

It is small only if most of the mass of $|f|^2$ is concentrated near the mean E. The behavior of the Fourier transform under scaling (Exercise 5 of Section 14H) suggests that if f has small variance, then $\widehat{f}$ may be expected to have large variance. Note that we need a factor of $1/2\pi$ to make the Fourier transform have norm 1, so the mean and variance for $\widehat{f}$ are

$$\widehat{E} = \frac{1}{2\pi} \int_{\mathbb{R}} \xi |\widehat{f}(\xi)|^2 d\xi; \qquad (52)$$

$$\widehat{V} = \frac{1}{2\pi} \int_{\mathbb{R}} (\xi - \widehat{E})^2 |\widehat{f}(\xi)|^2 d\xi. \qquad (53)$$

The remark above about the relation between the variances V and $\widehat{V}$ can be given quantitative form.

Proposition 14.6. *If f belongs to $L^2(\mathbb{R})$ such that $\|f\| = 1$, then the product of the variances of f and of $\widehat{f}$, $V \cdot \widehat{V}$, is at least $\frac{1}{4}$.*

Sketch of proof: We use the results of some of the exercises for Section 13H. Let Q and P be the linear transformations on the functions

$$Qf(x) = xf(x); \qquad Pf(x) = \frac{1}{i}\frac{df}{dx}(x).$$

Then the Fourier transform of Pf is $\xi \widehat{f}(\xi)$, so

$$V = \|(Q - E)f\|^2; \qquad \widehat{V} = \|(P - \widehat{E})\widehat{f}\|^2. \qquad (54)$$

Denoting the identity operator by I, note that

$$PQ - QP = -iI, \qquad (Qf, g) = (f, Qg), \qquad (Pf, g) = (f, Pg). \qquad (55)$$

It follows from (29) and the Cauchy-Schwartz inequality that

$$1 = \|f\|^2 = (f, f) = i(PQf - QPf, f) = i[(Qf, Pf) - (Pf, Qf)]$$
$$= 2\operatorname{Im}(Pf, Qf) \leq 2\|Qf\| \cdot \|Pf\|. \qquad (56)$$

Now it is also true that

$$(P - \widehat{E}I)(Q - EI) - (Q - EI)(P - \widehat{E}I) = -iI,$$

so we may repeat the calculation (56) with $Q - EI$ in place of Q and $P - \widehat{E}I$ in place of P to obtain the desired inequality. □

The simplest case in quantum mechanics consists of a single particle in one dimension. Its *wave function* is an element $\psi \in L^2(\mathbb{R})$ having norm 1. Any physical measurement is characterized by a linear transformation T defined on some subspace of $L^2(\mathbb{R})$ that has the property $(Tf, f) = (f, Tf)$ for all f in its domain. The theory is probabilistic: If the wave function of the particle at a given moment is ψ, then the mean and variance of the measurement of the quantity associated to T are

$$E_T = (T\psi, \psi); \qquad V_T = ||(T - E_T I)^2 \psi||^2. \tag{57}$$

In the usual representation of the wave function, the *position operator* is the operator Q above and the *momentum operator* is hP, where P is the operator above and the positive number h is Planck's constant. Thus the inequality proved above gives the quantitative form of the relationship between the uncertainty in measurement of the position and the uncertainty in measurement of the velocity, known as the *Heisenberg Uncertainty Principle*:

$$\sqrt{V_Q} \cdot \sqrt{V_{hP}} \geq \frac{h}{2}. \tag{58}$$

Exercises

1. Prove that the product on the left side of (58) is unchanged if ψ is replaced by $\sqrt{\rho}\psi(\rho x)$, for any fixed positive ρ. (Note that the new function also has norm 1.)
2. Prove that equality is obtained in (58) if ψ is the Gaussian function $\psi(x) = e^{-x^2/2}/\sqrt{2\pi}$. It follows from the preceding exercise that equality is also obtained by scaling this function.
3. Prove that equality is obtained in (58) *only* when ψ is one of the functions in the preceding exercise (possibly multiplied by a constant having modulus 1).
4. The *Schrödinger equation* for a free quantum mechanical particle in one dimension is

$$\frac{\partial u}{\partial t}(x, t) = ih \frac{\partial^2 u}{\partial x^2}(x, t), \qquad h \text{ a positive constant.}$$

   Discuss the solution of this equation for u periodic in x, given the value of $u(x, 0)$. Show that

$$\frac{1}{2\pi} \int_{-\pi}^{\pi} |u(x, t)|^2 \, dx \qquad \text{is constant, i.e., independent of } t.$$

5. Discuss the Schrödinger equation for functions that are not periodic.

15

Ordinary Differential Equations

At various points we have come across simple first-order and second-order ordinary differential equations. In this chapter we study equations and systems of arbitrary order and establish the basic existence, uniqueness, and representation theorems.

15A. Introduction

An ordinary differential equation (ODE) expresses, at each point of an interval that is the domain of some function u, a relationship between $u(x)$ and various derivatives $u^{(k)}(x)$. The equation is said to be *linear* if the relationship is linear. The *order* of the equation is the order of the maximum derivative that appears. Examples are

$$\text{(a) } [u'(x)]^2 + [u(x)]^2 = 1 \qquad \text{(b) } u''(x) = \sin u(x) \qquad (1)$$
$$\text{(a) } u'(x) + \cos x \cdot u(x) = e^x \qquad \text{(b) } u''(x) + u'(x) - 2u(x) = e^x. \qquad (2)$$

Examples (1) are nonlinear (= not linear) and examples (2) are linear. Examples (a) are of order 1 and examples (b) are of order 2. For reasons that should be clear, of the two linear equations, (2b) is said to have *constant coefficients*.

A *system* of ODEs consists of more than one equation for more than one function, for example,

$$u'(x) = v(x), \qquad v'(x) = \sin u(x). \qquad (3)$$

This system is said to be of order 1 (or of first order). The reader may note that the single equation (1b) is equivalent to the system (3) and that, similarly, one can find a first-order system of two equations for two functions that is equivalent to (2b). In fact, any ODE or system of ODEs is equivalent to some first-order system.

In this chapter we treat some basic but general topics in the existence and uniqueness of ODEs and systems of ODEs.

Exercises

1. Show that (3) is equivalent to (1b).
2. Find a first-order linear system of two equations that is equivalent to (2b).
3. (Gronwall's inequality) Suppose that f is a positive, continuously differentiable function on $[0, \infty)$, and suppose that there is a constant C such that $|f'(x)| \leq Cf(x)$ for all $x \geq 0$. Prove that $f(x) \leq e^{Cx} f(0)$ for all $x \geq 0$.

15B. Homogeneous Linear Equations

It is convenient once again to consider complex-valued functions u. As before, differentiability means differentiability of the real and imaginary parts, and $u' = (\operatorname{Re} u)' + i(\operatorname{Im} u)'$. Higher order derivatives u'', u''', ... are defined accordingly. We consider here the set of all *infinitely differentiable* functions

$$C^\infty(\mathbb{R}) = \{u : \mathbb{R} \to \mathbb{C} : \text{each derivative of } u \text{ exists and is continuous on } \mathbb{R}\}.$$

This is a vector space with the usual operations for functions. The derivative $D = d/dx$ and its iterates

$$Du = u', \qquad D^2 u = u'', \qquad D^3 u = u''', \ldots,$$

are linear transformations from $C^\infty(\mathbb{R})$ to itself. It is convenient to define the zeroth power of D as the identity operator I:

$$D^0 = I, \qquad D^0 u = Iu = u.$$

Given any polynomial with complex coefficients, such as

$$p(\lambda) = \lambda^n + a_{n-1}\lambda^{n-1} + \cdots + a_1\lambda + a_0, \qquad a_j \in \mathbb{C}, \tag{4}$$

there is a corresponding linear differential operator with complex coefficients

$$p(D) = D^n + a_{n-1}D^{n-1} + \cdots + a_1 D + a_0 I. \tag{5}$$

Note that these operators satisfy the same algebraic laws as polynomials: If p and q are polynomials, then

$$[p+q](D) = p(D) + q(D), \qquad [pq](D) = p(D)q(D). \tag{6}$$

(In algebraic terminology, the mapping from polynomials to differential operators is a homomorphism from the ring of polynomials to the ring of linear operators from $C^\infty(\mathbb{R})$ to itself.)

The main result in this section is a description of all the solutions of the equation $p(D)u = 0$, that is, the null space of the linear transformation $p(D)$. This is necessarily a subspace of $C^\infty(\mathbb{R})$. We show that it is a finite-dimensional subspace and find a basis.

Theorem 15.1. *Suppose that the polynomial (4) has the factorization*

$$p(\lambda) = \prod_{j=1}^{m}(\lambda - \lambda_j)^{d_j}, \qquad \lambda_j \text{ distinct.}$$

Then the solutions of the equation $p(D)u = 0$ are precisely the linear combinations of the n functions

$$x^k e^{\lambda_j x}, \qquad 0 \le k < d_j, \ 1 \le j \le m. \tag{7}$$

The proof takes a number of steps.

Lemma 15.2. $D^k u = 0$ *if and only if u is a polynomial (in x) of degree less than k.*

Proof: This is true for $k = 1$, by the Mean Value Theorem (applied to the real and imaginary parts), or by the Fundamental Theorem of Calculus. Induce on k: If $D^{k+1}u = D^k(Du) = 0$, then Du is a polynomial of degree less than k. There is a polynomial p of degree at most k such that $Dp = Du$; then $u - p$ is constant. □

Lemma 15.3. *If μ is a complex number, then $(D - \mu I)^k u = 0$ if and only if u has the form*

$$u(x) = p(x) e^{\mu x},$$

where p is a polynomial in x of degree less than k.

Proof: Write $p(x) = e^{-\mu x} u(x)$, so that $u(x) = e^{\mu x} p(x)$. By Leibniz's rule

$$(D - \mu I)(e^{\mu x} p(x)) = \mu e^{\mu x} p(x) + e^{\mu x} Dp(x) - \mu e^{\mu x} p(x) = e^{\mu x} Dp(x).$$

Repeating this calculation k times, we get

$$(D - \mu I)^k (e^{\mu x} p(x)) = \cdots = e^{\mu x} D^k p(x).$$

Lemma 15.2 implies that these expressions vanish if and only if p is a polynomial of degree less than k. □

Lemma 15.4. *Suppose that $q_1, q_2, \ldots, q_m$ are nonzero polynomials that have no root in common. Then there are polynomials $p_1, p_2, \ldots, p_m$ such that*

$$p_1(\lambda)q_1(\lambda) + p_2(\lambda)q_2(\lambda) + \cdots + p_m(\lambda)q_m(\lambda) \equiv 1. \tag{8}$$

Proof: Let $\mathcal{I}$ be the set whose elements are all polynomials that have the form of the left side of (8) for some choice of the polynomials p_j. For example, q_1 belongs to $\mathcal{I}$. If q_1 itself has degree 0, then it is constant and a multiple of it is 1. Otherwise, there are at least two polynomials in our original set $\{q_j\}$, and q_1 cannot divide every one of them. (If it did, then any root of q_1 would be common root of all.) Suppose that q_1 does not divide q_2. We may divide q_2 by q_1 to obtain $q_2 = s_1 q_1 + r_1$, where s_1 and r_1 are polynomials; the degree of the remainder term r_1 is less than the of q_1, denoted $\deg(q_1)$; and $r_1 \neq 0$. Now $r_1 = q_1 - sq_2$, so r_1 belongs to $\mathcal{I}$. If r_1 has degree 0, then 1 is a multiple and we are finished. Otherwise, r_1 does not divide some q_j and we may repeat the process: $q_j = s_2 r_1 + r_2$ with $\deg(r_2) < \deg(r_1)$, and $r_2 \neq 0$. Again

$$r_2 = q_j - s_2 r_1 = q_j - s_2(q_1 - s_1 q_2) = q_j - s_2 q_1 + (s_2 s_1) q_2$$

belongs to $\mathcal{I}$. This process of dividing and obtaining as remainder an element of $\mathcal{I}$ having lower degree can be continued until we reach an element $r \neq 0$ that divides all the q_j's and therefore has degree 0. (Again, any root of r would be a common root of the q_j's). □

Proof of Theorem 15.1: Define polynomials

$$q_j(\lambda) = \prod_{k \neq j} (\lambda - \lambda_k)^{d_k} = \frac{p(\lambda)}{(\lambda - \lambda_j)^{d_j}}. \tag{9}$$

These polynomials have no common roots, so we may find polynomials p_j to satisfy (8). There is a corresponding identity for the differential operators defined by these polynomials:

$$I = p_1(D)q_1(D) + p_2(D)q_2(D) + \cdots + p_m(D)q_m(D). \tag{10}$$

Now suppose that $p(D)u = 0$. Use (10) to write

$$u = \sum_{j=1}^{m} p_j(D)q_j(D)u = \sum_{j=1}^{m} q_j(D)p_j(D)u = \sum_{j=1}^{m} u_j. \tag{11}$$

Then

$$(D - \lambda_j)^{d_j} u_j = (D - \lambda_j)^{d_j} q_j(D) p_j(D) u = p(D) p_j(D) u = p_j(D) p(D) u = 0.$$

By Lemma 2, u_j is a linear combination of the functions (7), so the proof is complete. □

Remark. Theorem 15.1 shows that the dimension of the null space of $p(D)$ is at most equal to $\deg(p)$. It can be shown that the functions (7) are linearly independent, so the dimension is exactly $\deg(p)$; see Exercise 12.

Exercises

1. Suppose that p is a polynomial of degree 2. Show that for any constants c_0 and c_1, there is a unique solution to the problem
$$p(D)u = 0, \quad u(0) = c_0, \quad u'(0) = c_1.$$

2. Suppose that p has degree 2. Prove or disprove: For every choice of constants c_0, c_1, and μ, the problem
$$p(D)u = \mu u, \quad u(0) = c_0, \quad u(1) = c_1$$
has a unique solution.

3. Suppose that p is any polynomial and μ is any complex constant. Prove that $p(D)(e^{\mu x}) = p(\mu)e^{\mu x}$.

4. Find *all* the solutions of the problem
$$u''(x) + 2u'(x) + u(x) = e^x.$$

5. For what values of $\lambda \in \mathbb{C}$ (if any) does the problem
$$u''(x) - 2u'(x) + \lambda u(x) = 0, \quad u(0) = 1, \quad u(1) = 0$$
have a solution?

6. Let $u(t)$ denote the vertical displacement from equilibrium of a weight suspended from a spring. The motion, if the spring is frictionless and there is no air resistance, is described by the equation of motion (a) below, while the presence of friction gives equation (b), where K and F are positive constants. In each case discuss the behavior of $u(t)$ as $t \to +\infty$.

    (a) $u''(t) = -Ku(t).$ (b) $u''(t) = -Ku(t) - Fu'(t).$

7. Suppose that L is the differential operator $D^2 + bD + cI$. Find the necessary and sufficient conditions on the coefficients b and c so that *every* solution of the equation $Lu = 0$ has the property

    (a) $\lim_{x \to +\infty} |u(x)| = 0$ or (b) $\limsup_{x \to +\infty} |u(x)| < \infty.$

8. Show that the following subspace of $\mathcal{F}$,
$$\mathcal{F}_{r,k} = \operatorname{span}\{x^j e^{rx}; 0 \le j \le k\},$$
is invariant under D and therefore under each $p(D)$. Show that if p can be factored, then all solutions of $p(D)u = f$ can be found explicitly whenever f belongs to $\mathcal{F}_{r,k}$.

9. Suppose that g is a continuous function that vanishes for $|x| \geq M$ and suppose that u satisfies one of the following equations. What can you say about the behavior of u as $x \to +\infty$?

   (a) $D^2 u - Du + 2u = g$.   (b) $D^2 u + 2Du + 2u = g$.

10. Let $L = D^2 - I$. A *fundamental solution* for L is a continuous real- or complex-valued function G defined on $\mathbb{R}$ that has the properties

    (i) $G'(x)$ and $G''(x)$ exist for $x \neq 0$, and $G''(x) = G(x)$;

    (ii) $\lim_{\varepsilon \to 0, \varepsilon > 0} [G'(\varepsilon) - G'(-\varepsilon)] = 1$.

    (a) Find a *bounded* fundamental solution G for L.
    (b) Show that if g is a continuous function that vanishes for $|x| \geq M$, then the function
    $$u(x) = \int_{-\infty}^{+\infty} G(x - y) g(y) \, dy$$
    satisfies the equation $u'' - u = g$.

11. Suppose that the polynomial p in Theorem 15.1 has real coefficients. (Recall that this implies that the roots come in complex conjugate pairs.) Show that any real solution of $p(D)f = 0$ is a real linear combination of functions of the form
    $$x^k e^{r_j x} \cos s_j x, \qquad x^k e^{r_j x} \sin s_j x,$$
    where the numbers $\{r_j + i s_j\}$ are the roots of p.

12. Prove that the functions (7) are linearly independent.

13. The operator T defined by $Tu(x) = xu'(x)$ is a linear transformation from the space $C^\infty(\mathbb{R}_+)$ of infinitely differentiable functions on the half-line $\mathbb{R}_+ = (0, +\infty)$ to itself.
    (a) Find all solutions of $Tu = ru$, for constant $r \in \mathbb{C}$.
    (b) Find all solutions of $T^2 u + 2bTu + cu = 0$, at least in the case when the polynomial $\lambda^2 + 2b\lambda + c$ has distinct roots.

15C. Constant Coefficient First-Order Systems

A *first-order linear system* of ODEs is a system of equations of the form

$$u_1'(x) = a_{11}(x)u_1(x) + a_{12}(x)u_2(x) + \cdots + a_{1n}(x)u_n(x) + f_1(x),$$
$$u_2'(x) = a_{21}(x)u_1(x) + a_{22}(x)u_2(x) + \cdots + a_{2n}(x)u_n(x) + f_2(x),$$
$$\cdots$$
$$u_n'(x) = a_{n1}(x)u_1(x) + a_{n2}(x)u_2(x) + \cdots + a_{nn}(x)u_n(x) + f_n(x).$$

Using vector and matrix notation, we can write this much more concisely as

$$\mathbf{u}'(x) = A(x)\mathbf{u}(x) + \mathbf{f}(x),$$

where **u** and **f** are functions from $\mathbb{R}$ to $\mathbb{C}^n$ and $A(x)$ is an $n \times n$ matrix. The system is said to have *constant coefficients* if $A(x) = A$ is constant and is said to be *homogeneous* if $\mathbf{f} = 0$.

We consider here the constant coefficient $n \times n$ case and begin with the homogeneous system

$$\mathbf{u}'(x) = A\mathbf{u}(x). \tag{11}$$

Here A is fixed matrix. One way to look for a solution is to *assume* that the solution is given by a power series in the variable x. This would mean

$$\mathbf{u}(x) = \sum_{k=0}^{\infty} x^k \mathbf{u}_k, \tag{12}$$

where each $\mathbf{u}_k$ is a vector. Since convergent power series can be differentiated term by term, (11) and (12) imply

$$\sum_{k=0}^{\infty} k x^{k-1} \mathbf{u}_k = \sum_{k=0}^{\infty} x^k A \mathbf{u}_k. \tag{13}$$

Coefficients of x^k must be equal, so $\mathbf{u}_k = k^{-1} A \mathbf{u}_{k-1}$. Iterating, we can conclude that $\mathbf{u}_k = (k!)^{-1} A^k \mathbf{u}_0$. Now $\mathbf{u}_0 = \mathbf{u}(0)$, so we have derived the solution

$$\mathbf{u}(x) = \left[\sum_{k=0}^{\infty} \frac{1}{k!}(xA)^k\right] \mathbf{u}(0) = e^{xA} \mathbf{u}(0), \tag{14}$$

where we use the power series to *define* the matrix e^{xA}. Our derivation was rather formal, but the power series can be shown to converge; moreover, it can be differentiated term by term and so, in fact, it does provide a solution to (11). But, conversely, *every* solution has this form. To see this, suppose that **u** is a solution to (11) and note that Leibniz's rule implies

$$\frac{d}{dx}\left[e^{-xA} \mathbf{u}(x)\right] = -A e^{-xA} \mathbf{u}(x) + e^{-xA} A\mathbf{u}(x) = 0, \tag{15}$$

since A commutes with e^{-xA}. Therefore, $e^{-xA}\mathbf{u}(x)$ is constant; its value at $x = 0$ is $\mathbf{u}(0)$. Similarly, a differentiation shows that $e^{-xA} e^{xA}$ is constant, and its value at $x = 0$ is I, so **u** has the form (14). (Various loose ends in this argument are dealt with in the exercises.)

We have sketched a proof of the following.

Theorem 15.5. *Every solution of* $\mathbf{u}' = A\mathbf{u}$ *has the form* $\mathbf{u}(x) = e^{xA} \mathbf{u}_0$, *where* $\mathbf{u}_0$ *belongs to* $\mathbb{C}^n$. *Conversely, every function having the form* $e^{xA}\mathbf{u}_0$ *is a solution of* $\mathbf{u}' = A\mathbf{u}$.

Another proof can be given based on the result in Section 15E; see Section 15F.

Corollary 15.6. *The set of solutions of* $\mathbf{u}' = A\mathbf{u}$, *where A is $n \times n$, has dimension n.*

Consider now the inhomogeneous problem: Given the function $\mathbf{f}$, assumed to be continuous, determine $\mathbf{u}$ such that

$$\mathbf{u}'(x) = A\mathbf{u}(x) + \mathbf{f}(x), \qquad \mathbf{u}(0) = \mathbf{v}. \tag{16}$$

If $\mathbf{u}$ were a solution of (16), then a calculation like (15) would imply that

$$\frac{d}{dx}\left[e^{-xA}\mathbf{u}(x)\right] = e^{-xA}\mathbf{f}(x). \tag{17}$$

Therefore, integration from $x = 0$, followed by application of e^{xA}, gives

$$\mathbf{u}(x) = e^{xA}\mathbf{v} + e^{xA}\int_0^x e^{-yA}\mathbf{f}(y)dy = e^{xA}\mathbf{v} + \int_0^x e^{(x-y)A}\mathbf{f}(y)dy. \tag{18}$$

(At the last step we used the identity $e^{xA}e^{-yA} = e^{(x-y)A}$.)

The results in the preceding section and in this section, to this point, almost exhaust the cases for which a differential equation or system of a reasonably general type has a solution that can be expressed simply by a formula. We digress from the main topic of this section to mention the principal remaining cases. First are equations of the form $g(u(x))u'(x) = f(x)$, where g and u are known. Choose G so that $G' = g$ and let $v(x) = G(u(x))$. Then the differential equation for u becomes the equation $v'(x) = f(x)$. Second, consider the general first-order linear equation for a single function

$$u'(x) = a(x)u(x) + f(x). \tag{19}$$

Choose b so that $b' = a$. Then (19) is equivalent to the equation

$$\frac{d}{dx}\left[e^{-b(x)}u(x)\right] = e^{-b(x)}f(x), \tag{20}$$

whose general solution looks like

$$u(x) = e^{b(x)-b(0)}u(0) + \int_0^x e^{b(x)-b(y)}f(y)dy. \tag{21}$$

Some other cases in which the solution is (nearly) given by formulas are dealt with in Exercises 10 and 12.

Exercises

1. Write each of the following problems as a first-order system and use the matrix method to compute the solution:

    (a) $\quad u'' + u = 0, \quad u(0) = 0, \quad u'(0) = 1$

    (b) $\quad u'' - u = 0, \quad u(0) = 0, \quad u'(0) = 1$

    (c) $\quad u'' - 2u' + u = 0, \quad u(0) = 0, \quad u'(0) = 1.$

2. Fill in the details in the derivation of equations (17) and (18).
3. Find all solutions of the equation $u'(x) = xu(x) + e^x$.
4. Prove Corollary 15.6.
5. Suppose that B is an invertible matrix whose columns are eigenvectors of A. Show that the matrix $B^{-1}AB$ is diagonal. Find the system of equations for $v(x) = B^{-1}u(x)$ that corresponds to the system $u'(x) = Au(x)$. Apply these ideas to give a second derivation of a solution to Exercise 1(a).
6. Suppose that A is a real $n \times n$ matrix. Find the necessary and sufficient conditions on A such that every solution $\mathbf{u} : \mathbb{R} \to \mathbb{R}^n$ of the system $\mathbf{u}'(x) = A\mathbf{u}(x)$ has the property that the Euclidean norm $\|\mathbf{u}(x)\| = [\sum u_j(x)^2]^{1/2}$ is constant.
7. Define a norm on the space of $n \times n$ matrices by setting

$$\|A\| = \sup_{\|\mathbf{u}\| \leq 1} \|A\mathbf{u}\|,$$

    where, for a vector $\mathbf{u}$, $\|\mathbf{u}\|$ denotes the Euclidean norm in $\mathbb{R}^n$, as in the previous exercise.
    (a) Show that, always $\|A\mathbf{u}\| \leq \|A\| \cdot \|\mathbf{u}\|$.
    (b) Show that, for $n \times n$ matrices A and B, $\|AB\| \leq \|A\| \cdot \|B\|$. In particular, $\|A^k\| \leq \|A\|^k$.
    (c) Use (b) to show that the series defining e^A always converges.
8. Show that $e^A e^B = e^{A+B}$ if $AB = BA$.
9. Consider the system version of (19), in which $\mathbf{u}$ takes values in $\mathbb{C}^n$ and $a(x)$ is an $n \times n$ matrix. Show that (17) and (18) go through if the matrices $a(x)$, $a(y)$ commute for all x, y. Consider the question of what goes wrong if they do not commute.
10. A first-order equation $du/dx = f(x, u)$ is said to be *separable* if $f(x, u) = M(x)/N(u)$ for some functions m and N. Show that solutions are given implicitly by

$$\int M(x)\,dx = \int N(y)\,dy.$$

11. Find the solution to $du/dx = \cos x/2u$, $u(0) = 2$.
12. In the first-order equation $du/dx = f(x, u)$, assume that f satisfies the homogeneity condition: For all $t > 0$, $f(tx, t^a u) = t^{a-1} f(x, u)$. Show that the equation becomes separable (Exercise 10) in x and v if $v = x^{-a}u$.

15D. Nonuniqueness and Existence

The simplest nonlinear differential equation is the first-order equation for a real-valued function u:

$$u'(x) = f(x, u(x)),$$

where f is a given real function defined on an open subset Ω of $\mathbb{R}^2$. We assume that f is continuous. Geometrically, f should be thought of as specifying at each point of Ω a preferred direction or slope. A solution u of (1) is a function u such that, at each point of the graph u, the tangent line has the preferred direction. Intuitively it seems that there should be exactly one such maximal graph through each point of Ω, where maximal means that the graph is prolonged to the boundary of Ω. This intuition is half correct: Such a solution exists, but it may not be not unique. An example is

$$u'(x) = 3[u(x)]^{2/3}, \qquad u(0) = 0. \tag{22}$$

This has the following solutions: For any c and d such that $-\infty \le c \le 0 \le d < +\infty$,

$$u_{cd}(x) = \begin{cases} (x-c)^3, & x < c \\ 0, & c \le x \le d; \\ (x-d)^3, & x > d. \end{cases}$$

We shall see in the next section how to guarantee uniqueness. The proof of existence, in this generality, depends on an important result about compactness in a space of continuous functions. As in Chapter 7, if I is a bounded closed interval, $I = [a, b]$, then $C(I)$ denotes the space of continuous real-valued functions defined on I, with norm

$$\|u\|_{\sup} = \sup_{x \in I} |u(x)|.$$

Convergence in norm is the same as uniform convergence.

Definition. A collection $\mathcal{F}$ of functions in $C(I)$ is is said to be *equicontinuous* if for each $\varepsilon > 0$ there is $\delta > 0$ such that, for every $u \in \mathcal{F}$,

$$|u(x) - u(y)| < \varepsilon \quad \text{if } x, y \in I, \quad |x - y| < \delta.$$

(The important point here is that δ does not depend on u.)

Theorem 15.7: Theorem of Ascoli-Arzelà. *If $\mathcal{F}$ is a bounded, equicontinuous family of functions in $C(I)$, then every sequence in $\mathcal{F}$ contains a subsequence that converges in norm to an element of $C(I)$.*

Proof: Suppose that $(u_k)_{k=1}^\infty$ is a sequence in $\mathcal{F}$. For each positive integer n, partition the interval I into 2^n equal subintervals and partition the interval $[-M, M]$ into 2^n equal subintervals, where M is a bound for $\mathcal{F}$. This gives a partition of the rectangle $I \times [-M, M]$ into 4^n subrectangles. If u belongs to $\mathcal{F}$, then its graph is a subset of the large rectangle. Say that the *n-pattern* of u is the union of those of the 4^n subrectangles that are intersected by the graph of u. We choose a sequence of patterns as follows. At stage 1 there are nine possible patterns and at least one of them is the pattern of u_k for infinitely many values of k. Choose such a pattern P_1. Among the various possible 2-patterns that are subsets of P_1, choose one, P_2, that is the 2-pattern of u_k for *infinitely many* values of k, and continue. In this way we can choose a sequence of patterns and a subsequence of functions with

$$P_1 \supset P_2 \supset P_3 \supset \cdots, \qquad P_n \text{ is the } n\text{-pattern of } u_{k_n}.$$

Now we invoke the equicontinuity assumption. For any given $\varepsilon > 0$ there is an N so large that any $u \in \mathcal{F}$ varies by at most ε on intervals of length $2^{-N}|I|$; if we also choose N so that $2^{-N}M \leq \varepsilon$, it follows that any vertical slice of P_n has length $\leq 5\varepsilon$ for $n \geq N$. Thus

$$\|u_{k_n} - u_{k_m}\|_{\sup} \leq 5\varepsilon \qquad \text{if } n, m \geq N.$$

Therefore (since $C(I)$ is complete) the subsequence converges in norm to a function in $C(I)$. □

Remark. The theorem and the proof go through with little or no change if the interval I is replaced by a compact set in $\mathbb{R}^d$ and the functions take values in $\mathbb{C}^r$. In fact, the proof can be adapted to allow the domain and range to be any pair of compact metric spaces, using total boundedness (see Exercise 8 of Section 6D).

We are now in a position to prove the general existence theorem.

Theorem 15.8. Theorem of Peano. *Suppose that Ω is an open subset of $\mathbb{R}^2$ and suppose that f is a continuous real-valued function on Ω. For each point (x_0, y_0) in Ω, there is a continuously differentiable function u, defined in an open interval containing x_0, such that*

$$u'(x) = f(x, u(x)) \qquad \text{and} \qquad u(x_0) = y_0. \tag{23}$$

Proof: Since Ω is open and f is continuous, we can choose positive constants K and δ such that

$$\text{if } |x - x_0| \leq \delta \text{ and } |y - y_0| \leq K\delta, \text{ then } (x, y) \in \Omega \text{ and } |f(x, y)| \leq K. \tag{24}$$

Let $I = [x_0 - \delta, x_0 + \delta]$ and let $\mathcal{F}$ be the subset of $C(I)$ that consists of the functions u_k obtained as follows: Partition I into 2^k equal subintervals and take the unique continuous function whose graph goes through (x_0, y_0), has constant slope on each subinterval, and the slope on a subinterval is the value of f at the right endpoint of this portion of the graph if the subinterval is left of x_0, and at the left endpoint if the subinterval is to the right. Because of (24), these graphs stay in Ω. This family of functions is bounded and the limitation on slopes implies that it is equicontinuous, so some subsequence converges uniformly to a function $u \in C(I)$. Now each u_k is piecewise continuously differentiable and its derivative at a given point comes closer, *uniformly*, to the value of f at that point of the graph as k increases, because of continuity of f and the choice of the u_k. Therefore, for x in I,

$$u(x) = \lim_{n \to \infty} u_{k_n}(x) = \lim_{n \to \infty} \left[y_0 + \int_{x_0}^{x} u'_{k_n}(t)) \, dt \right] = y_0 + \int_{x_0}^{x} f(t, u(t)) \, dt.$$

This is equivalent to (23). □

Remark. This theorem is valid in much greater generality. For example, consider an n-th order equation

$$D^n u(x) = f(x, u(x), Du(x), \ldots, D^{n-1}u(x)), \tag{25}$$

where f is a continuous real-valued function defined on an open set $\Omega \subset \mathbb{R}^{n+1}$. This can be converted to a first-order system if we think of $u_k = D^{k-1}u$ for $1 \leq k < n$. The system is

$$u'_k(x) = u_{k+1}(x), \quad 1 \leq k < n;$$
$$u'_n(x) = f(x, u_1(x), u_2(x), \ldots, u_n(x)).$$

This is a special case of the system

$$\mathbf{u}'(x) = \mathbf{f}(x, \mathbf{u}(x)) \tag{26}$$

for a vector-valued function $\mathbf{u} : I \to \mathbb{R}^n$, where $\mathbf{f}$ is a continuous function from Ω to $\mathbb{R}^n$. Again, this system has a solution, not necessarily unique, through each point of Ω:

$$\mathbf{u}(x_0) = \mathbf{y}_0, \quad (x_0, \mathbf{y}_0) \in \Omega. \tag{27}$$

The proof is the same as before.

Going back to (25), condition (27) means specifying u and its derivatives of order less than n at a given point.

Remark. The fact that one may need to pass to a subsequence of the sequence of functions constructed in the proof of Peano's Theorem in order to get convergence means that, in this generality, the construction is of no practical value. However, if f satisfies the stronger hypothesis of the next section, then the u_k's themselves converge.

Exercises

1. With the example of nonuniqueness (22), explain how, starting at a point (x_0, y_0) with $y_0 \neq 0$, one might have to pass to a subsequence of the sequence in the proof of Peano's Theorem in order to get convergence.
2. Can a system of two second-order equations

$$u''(x) = f(x, u(x), v(x), u'(x), v'(x))$$
$$v''(x) = g(x, u(x), v(x), u'(x), v'(x))$$

be reduced to a first-order system?
3. Prove the assertion made in the proof of Peano's Theorem that the family of functions constructed there is an equicontinuous family.

15E. Existence and Uniqueness

In this section we show that a seemingly slight strengthening of the continuity hypothesis on the function f in (23) or $\mathbf{f}$ in (26) leads both to uniqueness and to an efficient method of approximating solutions. We consider the vector-valued version of the problem (23), (24) and note that these together are equivalent to the vector-valued *integral equation*

$$\mathbf{u}(x) = \mathbf{y}_0 + \int_{x_0}^{x} \mathbf{f}(t, \mathbf{u}(t)) \, dt. \tag{28}$$

Picard's iteration method or *method of successive approximations* for solving (1) is to make an initial guess at a solution and then refine the guess by using (28). One obtains the sequence of vector-valued functions known as the *Picard iterates* for (28):

$$\mathbf{u}_0(x) \equiv \mathbf{y}_0; \quad \mathbf{u}_{k+1}(x) = \mathbf{y}_0 + \int_{x_0}^{x} \mathbf{f}(t, \mathbf{u}_k(t)) \, dt, \quad k = 0, 1, 2, \ldots. \tag{29}$$

If these iterates converge uniformly on some interval I containing x_0, then the limit $\mathbf{u}$ satisfies (28) and is therefore a solution of (26), (27).

Another point of view suggested by (28) is to view our problem as a *fixed point* problem. Suppose for the sake of simplicity that the domain of definition Ω of F is all of $\mathbb{R}^{n+1}$. Given any bounded closed interval I, let $C(I)$ now denote the space of vector-valued continuous functions

$$C(I) = \{\mathbf{u} : I \to \mathbb{R}^n; \ \mathbf{u} \text{ is continuous}\}; \quad \|\mathbf{u}\|_{\sup} = \sup_{x \in I} \|\mathbf{u}(x)\|. \tag{30}$$

15E. Existence and Uniqueness

Then define a mapping S from $C(I)$ to itself by

$$[S(\mathbf{u})](x) = \mathbf{y}_0 + \int_{x_0}^{x} \mathbf{f}(t, \mathbf{u}(t)) \, dt. \tag{31}$$

Then (28) says that we want a *fixed point* of S: an element $\mathbf{u} \in C(I)$ such that $\mathbf{u} = S(\mathbf{u})$. Now $C(I)$ is a complete metric space, so we can hope to make use of the following.

Banach Fixed Point Theorem. *Suppose that X is a nonempty complete metric space with metric d and suppose that the function S from X to itself is a strict contraction, that is, for some positive constant $\rho < 1$, and any x and y in S,*

$$d(S(x), S(y)) \leq \rho \, d(x, y). \tag{32}$$

Then S has a unique fixed point in X.

Proof: Choose any point $x_1 \in X$ and define a sequence by iteration: $x_{k+1} = S(x_k)$. Condition (32) implies that

$$d(x_{k+2}, x_{k+1}) \leq \rho \, d(x_{k+1}, x_k) \leq \rho^2 d(x_k, x_{k-1}) \leq \cdots \leq \rho^k d(x_2, x_1).$$

Therefore

$$d(x_{k+m}, x_{k+1}) \leq d(x_{k+m}, x_{k+m-1}) + d(x_{k+m-1}, x_{k+m-2}) + \cdots + d(x_{k+2}, x_{k+1})$$

$$\leq [\rho^{k+m-2} + \rho^{k+m-3} + \cdots + \rho^k] d(x_2, x_1) \leq \frac{\rho^k}{1-\rho} d(x_2, x_1).$$

It follows that $(x_k)_{k=1}^{\infty}$ is a Cauchy sequence in X, so it converges to a point $x \in X$. Now S is continuous, so

$$d(x, S(x)) = \lim_{k \to \infty} d(x_k, S(x_k)) = \lim_{k \to \infty} d(x_k, x_{k+1}) = 0.$$

If x' is also a fixed point, then

$$d(x, x') = d(S(x), S(x')) \leq \rho \, d(x, x'),$$

so $d(x, x') = 0$ and $x' = x$. □

Definition. A function $\mathbf{f}(x, \mathbf{y})$ defined for certain values of $x \in \mathbb{R}$ and $\mathbf{y} \in \mathbb{R}^n$ and having values in $\mathbb{R}^n$ is said to satisfy a *Lipschitz condition* with respect to $\mathbf{y}$ if there is a constant K such that

$$\|\mathbf{f}(x, \mathbf{y}') - \mathbf{f}(x, \mathbf{y})\| \leq K \, \|\mathbf{y} - \mathbf{y}'\| \tag{33}$$

whenever the left side is defined. (This will be true if the components of F have partial derivatives in the y-variables that are bounded on the domain of definition, provided the domain is, for example, a ball or rectangle.) The constant K is called a *Lipschitz constant*.

Theorem 15.9: Existence and Uniqueness Theorem. *Suppose that Ω is an open subset of $\mathbb{R}^{n+1}$, and suppose that $\mathbf{f}$ is a continuous function from Ω to $\mathbb{R}^n$ that satisfies a Lipschitz condition with respect to $\mathbf{y}$. Then, for each point $(x_0, \mathbf{y}_0)$ in Ω, the problem (28) has a solution on some open interval containing x_0, and any two such solutions coincide on their common domain of definition.*

Proof: Given $(x_0, \mathbf{y}_0) \in \Omega$, choose $\varepsilon > 0$ and $r > 0$ small enough that

$$\{(x, \mathbf{y}) : |x - x_0| \leq \varepsilon, \|\mathbf{y} - \mathbf{y}_0\| \leq r\} \subset \Omega.$$

Let

$$N = \sup_{|x - x_0| \leq \varepsilon} \|\mathbf{f}(x, \mathbf{y}_0)\|.$$

Let K be a Lipschitz constant for $\mathbf{f}$ and let $J = [x_0 - \delta, x_0 + \delta]$, where

$$\delta = \min\{\varepsilon, r/2N, r/2K\}.$$

Let $\mathbf{u}_0 \equiv \mathbf{y}_0$ be the first Picard iterate, and let S be the mapping defined by (4). Let $X \subset C(I)$ be the closed ball

$$X = \{\mathbf{u} \in C(I) : \|\mathbf{u} - \mathbf{u}_0\|_{\sup} \leq r\}.$$

By our choice of δ,

$$\|S(\mathbf{u}_0) - \mathbf{u}_0\|_{\sup} = \sup_{|x-x_0|\leq\delta} \left\| \int_{x_0}^{x} \mathbf{f}(t, \mathbf{y}_0) dt \right\| \leq \delta N \leq \frac{r}{2}. \tag{34}$$

If $\mathbf{u}$ and $\mathbf{v}$ belong to X, then $S(\mathbf{u})$ and $S(\mathbf{v})$ are defined, and an elementary argument using (33) shows that

$$\|S(\mathbf{u}) - S(\mathbf{v})\|_{\sup} \leq \delta K \|\mathbf{u} - \mathbf{v}\|_{\sup} \leq \tfrac{1}{2} \|\mathbf{u} - \mathbf{v}\|_{\sup}. \tag{35}$$

In particular, we can take $\mathbf{v} = \mathbf{u}_0$ in (35) and use (35) to conclude that S maps X to itself. The Banach Fixed Point Theorem gives us a unique solution to (28) in the ball X. □

The Picard iterates are precisely the sequence obtained by starting with $\mathbf{u}_0 = \mathbf{y}_0$ and $\mathbf{u}_{k+1} = S(\mathbf{u}_k)$. Therefore the proofs of the two preceding theorems together give the following.

15E. Existence and Uniqueness

Corollary 15.10. *Under the assumptions of the Existence and Uniqueness Theorem, the Picard iterates on an interval containing x_0 converge uniformly to a solution of (28).*

Remark. The interval of existence that is obtained in the proof of Theorem 15.9 is usually not as large as it could be. Suppose, for example, that the function $\mathbf{f}$ is defined and continuous for all $x \in \mathbb{R}$ and all $\mathbf{y} \in \mathbb{R}^n$ and satisfies (6) everywhere. The proof of Theorem 15.5 given above only proves convergence on a certain bounded interval, but in fact the Picard iterates converge on all of $\mathbb{R}$; see Exercise 8.

Exercises

1. Compute the first three Picard iterates for the problem
$$u'(x) = x + u(x)^2, \qquad u(0) = 1.$$

2. Compute the first three Picard approximations to the solution of the first-order *system* associated to the problem
$$u''(x) = u(x), \qquad u(0) = 1, \quad u'(0) = 0.$$

3. Suppose that A is an $n \times n$ matrix. Investigate the Picard iterates of the problem for $\mathbf{u} : \mathbb{R} \to \mathbb{C}^n$:
$$\mathbf{u}'(x) = A\mathbf{u}(x), \qquad \mathbf{u}(0) = \mathbf{v}.$$

4. Write the formula for the Picard iterates of the system associated to the problem
$$u''(x) + u(x) + \cos[x - u'(x)] = 0; \qquad u(0) = u'(0) = 0.$$

5. Discuss the question of existence and uniqueness of solutions to the following problems:

   (a) $u'(x) = (\sin^2[u(x) + x])^{1/3}$, $\qquad u(0) = 0$
   (b) $u'(x) = (1 + \sin^2[u(x) + x])^{1/3}$, $\qquad u(0) = 0$
   (c) $u''(x) = \cos u(x)$, $\qquad u(0) = u'(0) = 0$
   (d) $u''(x) = u(x)^2$, $\qquad u(0) = 1, u'(0) = 0$
   (e) $u''(x) = u(x)^{1/3}$, $\qquad u(0) = 0, u'(0) = 0.$

6. Prove the inequalities (34) and (35).
7. Prove Corollary 15.10 by showing first that there is a solution on some largest subinterval of J; if this subinterval were not all of J, then it would have an endpoint that belongs to J. What would happen near this endpoint?

8. Suppose that $\mathbf{f}(x, u)$ is defined for all real x and all $\mathbf{y}$ in $\mathbb{R}^n$ and satisfies

$$\|\mathbf{f}(x, \mathbf{y}') - \mathbf{f}(x, \mathbf{y})\| \leq K \|\mathbf{y}' - \mathbf{y}\|, \qquad \text{all } x, \mathbf{y}, \mathbf{y}'.$$

(a) Suppose that J is any bounded interval that contains x_0. Prove that the Picard iterates (30) satisfy

$$\|\mathbf{u}_n - \mathbf{u}_{n-1}\| \leq C \frac{K^n}{n!} |x - x_0|, \qquad x \in J, \quad n = 1, 2, \ldots,$$

where

$$C = \sup \{\|\mathbf{f}(x, \mathbf{u}_0)\| : x \in J\}.$$

(b) Deduce from this sequence of inequalities that the iterates converge uniformly on the interval J.
(c) Suppose that $\mathbf{u}$ and $\mathbf{v}$ are two solutions of (28) on the interval J. Deduce a sequence of similar inequalities for $\|\mathbf{v} - \mathbf{u}\|$ on the interval and use them to show that $\mathbf{v} = \mathbf{u}$ on J.

15F. Linear Equations and Systems, Revisited

An important application of the result of the previous section is to linear equations and systems with variable coefficients. For first-order systems this means a problem like

$$\mathbf{u}'(x) = A(x)\mathbf{u}(x) + \mathbf{f}(x); \qquad \mathbf{u}(x_0) = \mathbf{y}_0. \tag{36}$$

Again $\mathbf{u}$ takes values in $\mathbb{C}^n$ and $A(x)$ is an $n \times n$ matrix. Set

$$\mathbf{f}(x, \mathbf{y}) = A(x)\mathbf{y} + \mathbf{f}(x). \tag{37}$$

Then

$$\|\mathbf{f}(x, \mathbf{y}) - \mathbf{f}(x, \mathbf{y}')\| = \|A(x)[\mathbf{y} - \mathbf{y}']\| \leq \|A(x)\| \cdot \|\mathbf{y} - \mathbf{y}'\|, \tag{38}$$

where the matrix norm is defined by

$$\|A\| = \sup_{\|\mathbf{y}\| \leq 1} \|A\mathbf{y}\|.$$

(See Exercise 7 of Section 15C.)

Proposition 15.11. *Suppose that the vector-valued function* $\mathbf{f}$ *and the matrix-valued function* A *are defined and continuous on an interval* $J \subset \mathbb{R}$*. Then the function* $\mathbf{F}$ *defined by* (37) *is continuous. Moreover,* $\mathbf{F}$ *satisfies a Lipschitz condition on every closed, bounded subinterval* I*.*

15F. Linear Equations and Systems, Revisited

In fact, $\|A(x)\|$ depends continuously on x and is therefore bounded on I, so (38) implies the Lipschitz condition.

Corollary 15.12. *Under the same hypotheses, the problem (36) has a unique solution on the interval J for every $x_0 \in J$ and every $\mathbf{y}_0$.*

The existence of a unique solution on some subinterval around x_0 follows immediately from previous results. The fact that there is solution on the whole interval J is left as an exercise.

A linear equation of higher order

$$D^n u(x) + a_{n-1}(x) D^{n-1} u(x) + \cdots + a_0(x) u(x) = g(x), \tag{39}$$

$$u(x_0) = y_0, \quad Du(x_0) = y_1, \cdots D^{n-1} = y_{n-1} \tag{40}$$

can be reduced to a linear first-order system of the form (1). Continuity of the coefficients a_j and of the function g implies the continuity of A and of $\mathbf{f}$.

Corollary 15.13. *If the coefficients a_j and the function g are continuous on an interval J, then for each x_0 in J the problem (39), (40) has a unique solution on the interval J.*

Exercises

1. Suppose that the $n \times n$ matrix-valued function A is continuous on an interval J. Show that the set of solutions to $\mathbf{u}' = A\mathbf{u}$ is an n-dimensional subspace of the vector space of continuous functions from J to $\mathbb{C}^n$.
2. Deduce from Exercise 1 that the dimension of the space of solutions to a homogeneous linear equation of order n with continuous coefficients (i.e., (36) with $g = 0$) is n. Consequently, the solutions found in Section 15B in the constant coefficient case are linearly independent.
3. Show in detail how Corollary 15.13 follows from Corollary 15.12.
4. Suppose that b and c are constants. Prove that every solution of the differential equation $x^2 D^2 u + bx Du + cu = 0$ is a linear combination of powers x^r and (possibly) of terms of the form $x^s \log x$ for suitable choices of r and (if necessary) s. When is it necessary to use a term like $x^s \log x$?

Appendix
The Banach-Tarski Paradox

Here we give a brief sketch of the theorem stated in Section 10A. We start with a lemma whose proof is sketched in the exercises. It deals with the rotations in $\mathbb{R}^3$, that is, the group $SO(3)$ of 3×3 orthogonal matrices having determinant 1. Any subset of $SO(3)$ generates a subgroup; if the subset has two elements σ and τ, then the subgroup G consists of the identity matrix, which we denote by **1**, and all matrices of the form

$$\sigma^{j_1}\tau^{k_1}\sigma^{j_2}\tau^{k_2}\cdots\sigma^{j_n}\tau^{k_n}, \tag{A.1}$$

where each j_m and each k_m is an integer (possibly negative); we also require that each of these integers, except possibly j_1 and k_n be nonzero, and when $n = 1$ we require that at least one of j_1 and k_1 be nonzero. The group G is said to be *free* if there is only one such expression for each matrix in G; this is equivalent to saying that **1** cannot be expressed in the form (A.1) with the limitations we have placed on the exponents.

Lemma A.1. *There are elements σ, τ in $SO(3)$ with the property that the subgroup G that they generate is free.*

A proof of Lemma A.1 is sketched in the exercises at the end of this section.
From now on we take σ and τ to be elements that generate a free subgroup G. If H is a subset of G, we write σH for the subset $\{\sigma\mu : \mu \in H\}$, and so on. The following result shows that a subset of G can be cut into four pieces that can be reassembled, using only the rotations σ and τ, to give two copies of G itself.

Lemma A.2. *There are pairwise disjoint subsets H_1, H_2, H_3, H_4 of G, whose union does not contain **1**, with the property that*

$$G = \sigma H_1 \cup \tau H_2, \qquad \sigma H_1 \cap \tau H_2 = \emptyset;$$
$$G = \sigma^{-1} H_3 \cup \tau^{-1} H_4, \qquad \sigma^{-1} H_3 \cap \tau^{-1} H_4 = \emptyset.$$

Proof: Let $W(\sigma^{-1})$ denote the set of elements of G that begin with σ^{-1} in the expression (16.1), that is, those with $j_1 < 0$. Similarly, let $W(\tau^{-1}\sigma)$ consist of those elements for which $j_1 = 0, k_1 = -1, j_2 > 0$, and so on. Then G is the disjoint union of the sets $\{\mathbf{1}\}$, $W(\sigma)$, $W(\sigma^{-1})$, $W(\tau)$, $W(\tau^{-1})$. Let

$$H_1 = W(\sigma^{-1}), \qquad H_2 = W(\tau^{-1}\sigma), \qquad H_3 = W(\sigma), \qquad H_4 = W(\tau\sigma^{-1}).$$

These sets are pairwise disjoint and do not contain $\mathbf{1}$ or τ^2, for example. It is easy to check that

$$\sigma H_1 = \{\mathbf{1}\} \cup W(\sigma^{-1}) \cup W(\tau) \cup W(\tau^{-1}), \qquad \tau H_2 = W(\sigma),$$

and similarly for $\sigma^{-1} H_3$ and $\tau^{-1} H_4$. □

Now let $A = \{x \in \mathbb{R}^3 : |x| \leq 1\}$ be the closed unit ball in $\mathbb{R}^3$. Each element of G takes A to itself. The *orbit* of a point $x \in A$ is the set $Gx = \{\mu x : \mu \in G\}$. The origin 0 has only the trivial orbit $\{0\}$. Every other point of A can be shown to have a countable orbit (Exercise 6 below). A subset $X \subset A$ is said to be a *cross-section* for the action of G if every X contains exactly one point from each nontrivial orbit. According to the Axiom of Choice, there is a cross-section.

Lemma A.3. *There are pairwise disjoint subsets A_1, A_2, A_3, A_4 of the ball A, whose union is a proper subset of A, with the property that*

$$A = \sigma A_1 \cup \tau A_2, \qquad \sigma A_1 \cap \tau A_2 = \emptyset;$$
$$A = \{0\} \cup \sigma^{-1} A_3 \cup \tau^{-1} A_4, \qquad \sigma^{-1} A_3 \cap \tau^{-1} A_4 = \emptyset.$$

Proof: Let X be a cross-section, let the H_j be as in Lemma 2, let $H_j X = \{\mu x : \mu \in H_j, x \in X\}$, and set

$$A_1 = H_1 X \cup \{0\}, \qquad \tilde{A}_2 = H_2 X, \qquad \tilde{A}_3 = H_3 X, \qquad A_4 = H_4 X;$$
$$A_2 = \tilde{A}_2 \setminus \{\tau^{-1} \sigma A_1 \cap \tilde{A}_2\}; \qquad A_3 = \tilde{A}_3 \setminus \{\tau \sigma^{-1} \tilde{A}_3 \cap A_4\}.$$

Note that any point of X is not in the union of the A_j's. Since $A = GX \cup \{0\}$, the desired properties follow from Lemma A.2. □

Now let B and C be the closed unit balls with centers $b = (0, 0, 3)$ and $c = (0, 0, -3)$, respectively.

Lemma A.4. *There are pairwise disjoint sets B_1, B_2, C_1, C_2, C_3 and a function $f : B \cup C \to A$ with the properties:*

$$B = B_1 \cup B_2, \qquad C = C_1 \cup C_2 \cup C_3;$$

the sets $f(B_1)$, $f(B_2)$, $f(C_1)$, $f(C_2)$, $f(C_3)$ are pairwise disjoint; and the restriction of f to any of B_j or C_j is a distance-preserving mapping (congruence).

Proof: The map $x \to \sigma x + b$ takes the set A_1 of Lemma A.3 onto a subset B_1 of B, while the map $x \to \tau x + b$ takes A_2 onto a subset B_2 of B. By Lemma A.3, B_1 and B_2 are disjoint and have union B. We take $f : B \to A_1 \cup A_2$ to be the inverse of the map just defined. Similarly, the map $x \to \sigma^{-1} x + c$ takes A_3 to C_1 and the map $x \to \tau^{-1} x + c$ takes A_4 to C_2. To exhaust C we must take one extra point $a \in A$ that does not belong to any A_j and translate it to the center $c \in C$; then $C_3 = \{c\}$, and again $f : C \to A_3 \cup A_4 \cup \{a\}$ is the inverse of the mapping just defined. □

We can now prove the Banach-Tarski Theorem, which we restate here, in slightly different notation.

Theorem. *There are pairwise disjoint sets $A'_1, \ldots, A'_7, B'_1, B'_2, B'_3$, and $C'_1, \ldots C'_4$ such that*

$$A = A'_1 \cup \cdots \cup A'_7, \quad B = B'_1 \cup B'_2 \cup B'_3, \quad C = C'_1 \cup \cdots \cup C'_4;$$
$$A'_1 \cong B'_1, \quad A'_2 \cong B'_2, \quad A'_3 \cong B'_3, \quad A'_4 \cong C'_1, \quad A'_5 \cong C'_2, \quad A'_6 \cong C'_3, \quad A'_7 \cong C'_4.$$

Proof: Define $g : A \to B \cup C$ to be translation by b, so $g(A) = B$. Let $D = B \cup C$, so that (using Lemma A.4) we have the 1–1 mappings (piecewise congruences)

$$f : D \to A, \qquad g : A \to D.$$

Given a point $x \in A$, we say that x has an *ancestor* in D if $x = f(y)$ for some $y \in D$; similarly we say that $y \in D$ has an ancestor in A if $y = g(x)$ for some $x \in A$—which is true if and only if $y \in B$. Given any point in A or in D that has an ancestor, we can search for successive ancestors, that is, ancestors of the ancestors. Either this process terminates (for example, it terminates immediately in D for $y \in C$) or it continues without terminating. This allows us to partition each of A and D into three subsets:

$$A_f = \{x \in A : \text{the ancestor search terminates in } D\}$$
$$A_g = \{x \in A : \text{the ancestor search terminates in } A\}$$
$$A_\infty = \{x \in A : \text{the ancestor search does not terminate}\},$$

and similarly D_f, D_g, and D_∞. It is clear that $f : D_f \to A_f$ is 1–1 and onto (bijective); similarly $g : A_g \to D_g$ is 1–1 and onto, while also $f : D_\infty \to A_\infty$ and $g : A_\infty \to D_\infty$ are inverses of each other. It follows from this and Lemma A.4 that we can complete the proof by setting

$$B'_j = B_j \cap (D_f \cup D_\infty), \quad j = 1, 2; \qquad B'_3 = B \cap D_g;$$
$$C'_j = C_j \cap (D_f \cup D_\infty), \quad j = 1, 2, 3; \qquad C'_4 = C \cap D_g;$$
$$A'_j = f(B'_j), \quad j = 1, 2; \qquad A'_3 = g^{-1}(B'_3);$$
$$A'_{j+3} = f(C'_j), \quad j = 1, 2, 3; \qquad A'_7 = g^{-1}(C'_4). \qquad \square$$

Exercises

1. Rotation by an angle θ in $\mathbb{R}^2$ is represented by a 2×2 orthogonal matrix

$$\mu = \begin{bmatrix} \cos\theta & -\sin\theta \\ \sin\theta & \cos\theta \end{bmatrix}.$$

   Check by matrix multiplication that rotation by θ_1 followed by rotation by θ_2 is rotation by $\theta_1 + \theta_2$. In particular, this means that the entries of μ^n are $\cos n\theta$ and $\pm \sin n\theta$.

2. Suppose in Exercise 1 that $\cos\theta = \alpha$ and $\sin\theta = \beta$. Then $e^{i\theta} = \alpha + i\beta$ and it follows that $\cos n\theta$ and $\sin n\theta$ are the real and imaginary parts of $e^{in\theta} = (\alpha + i\beta)^n$. Deduce that

$$\cos n\theta = \sum_k \binom{n}{2k}(-1)^k \alpha^{n-2k} \beta^{2k}; \qquad \sin n\theta = \sum_k \binom{n}{2k+1}(-1)^k \alpha^{n-2k-1} \beta^{2k+1}.$$

3. In Exercise 2, suppose that $\alpha = 1/3$ and $\beta = 2\sqrt{2}/3$. Show that

$$\cos n\theta = \left[\sum_k \binom{n}{2k}(-8)^k\right]\frac{1}{3^n}; \qquad \sin n\theta = \left[\sum_k \binom{n}{2k+1}(-8)^k\right]\frac{2\sqrt{2}}{3^n}.$$

Note that the identities $(1+1)^n = 2^n$ and $(1-1)^n = 0$ imply that

$$\sum_k \binom{n}{2k} = \sum_k \binom{n}{2k+1} = 2^{n-1}.$$

Use these identities and the fact that $-8 \equiv 1 \pmod{3}$ to deduce that

$\cos n\theta = \alpha_n/3^n$, $\sin n\theta = \beta_n 2\sqrt{2}/3^n$, where α_n, β_n are integers not divisible by 3.

4. With α and β as in Exercise 3, let σ and τ be the 3×3 orthogonal matrices

$$\sigma = \begin{bmatrix} \alpha & -\beta & 0 \\ \beta & \alpha & 0 \\ 0 & 0 & 1 \end{bmatrix}; \qquad \tau = \begin{bmatrix} 1 & 0 & 0 \\ 0 & \alpha & -\beta \\ 0 & \beta & \alpha \end{bmatrix}.$$

Suppose that μ is the element of the group G generated by σ and τ that has the form (A.1). Prove (by induction on N) that if $k_n \neq 0$, then the vector

$$\mu \begin{bmatrix} 0 \\ 0 \\ 1 \end{bmatrix} = 3^{-N} \begin{bmatrix} a \\ b\sqrt{2} \\ c \end{bmatrix},$$

where a, b, and c are integers and $N = \sum |j_m| + \sum |k_m|$. Moreover, show that c is divisible by 3 if $j_1 \neq 0$ while a is divisible by 3 if $j_1 = 0$ and $k_1 \neq 0$.

5. Use the results of Exercises 3 and 4 to prove that the group G generated by σ and τ is free. In fact, show that an element μ of form (A.1) with $k_n \neq 0$ cannot be the identity matrix by showing (by induction) that the vector in Exercise 4 has a middle entry $b\sqrt{2}/3^N$, where b is an integer not divisible by 3 and so $b \neq 0$. On the other hand, if μ has the form (A.1) with $k_n = 0$ but $j_n \neq 0$, then necessarily $k_{n-1} \neq 0$ and $\mu = 1$ if and only if $\mu' = \sigma^{j_n}\mu\sigma^{-j_n} - 1$.

6. Show that an orbit of G on A distinct from $\{0\}$ is infinite. [Hint: If x is in a finite orbit, show that there are positive integers m, n such that $\sigma^m x = x = \tau^n x$ and deduce from this that σ^m and τ^n must commute.]

Hints for Some Exercises

Section 1A

3. Multiply by $1 - r$.
4, 5. Write the terms in the n-th expression in terms of n and put over a common denominator.

Section 1B

5. Add $-z$ to both sides of $z + 0 = z + 0'$.
6. Add $-m$ to both sides; show that the resulting expression for x is the desired solution.
11. Multiply the identity $0 + 0 = 0$ by r and use the result of Exercise 5 or 6.
12. Multiply by r^{-1}.
13, 14. Adapt the divisibility argument used for $r^2 = 2$.

Section 1C

5. $S + 0^* = S$
6. For s to belong to $-S$, one must have $r + s \in 0^*$ for every $r \in S$.
8. Hint: What positive rationals should the product contain? What other rationals?

Section 2A

1. Convert this statement so that O5 applies.
2. Start with a rational $r_0 < x$ and an irrational $t_0 < x$, and add multiples of a sufficiently small rational.
3. Can the set consisting of positive integer multiples of ε be bounded?
5. The set $\{a_1, a_2, \ldots\}$ has a least upper bound a; show that a is the desired (unique) point.

6. Suppose that A is nonempty and bounded above. Show that there is some integer M such that $M + 1$ is an upper bound and n is not; let $a_1 = M, b_1 = M + 1$. Then, among the three numbers $M, M + \frac{1}{2}, M + 1$, we can choose a_2 and b_2 so that a_2 is not an upper bound, b_2 is an upper bound, and $b_2 - a_2 = \frac{1}{2}$. Continue.
7. (a) Show that if $h > 0$ is small enough, then $(x + h)^2 < 2$, and then use Exercise 2.
   (b) Adapt the method in (a).

9–11. Induction.

12. Write $a = 1 + h$, so $(1 + h)^n = n$, and use Exercise 11.
13. See Exercise 12.
14. True. We can suppose that m is very large and let $n = m + k$; then $\sqrt{n} - \sqrt{m} \approx \pi$ implies $k = n - m \approx \pi(\sqrt{n} + \sqrt{m}) < 2\pi\sqrt{m}$, which implies that k is much smaller than m, so it is reasonable to try taking k close to $2\pi\sqrt{m}$.

15, 16. Divisibility.

17. Uniqueness: What if $x < y$?
18. This amounts to finding a (presumed) inequality and checking whether it is necessarily true.

Section 2B

1. Choose the q_j's as large as possible at each step.
2. The numbers with a 7 in the first decimal place make up an interval of length $1/10$ in $[0, 1]$, so the total length of the remaining intervals is $9/10$. What is the total length of the intervals corresponding to no 7 in the first two decimal places?

Section 2C

1. Look at the dimensions of the appropriate vector spaces.

Section 2D

1. What happens to successive sums of 1?
2. Should one have $i < 0$ or $i > 0$?
5. Suppose that z has the given form. What do you learn from looking at the product $z\bar{z}$?
7. Interpret this as saying two distances are equal.
8. The relations do not change if we rotate the plane, so we can assume that $a = 1$. Then necessarily $c = \bar{b} = b^{-1}$.

15, 16. Use Exercise 14.

Section 3A

1. It is enough to prove that, for every $\varepsilon > 0$, it follows that $a - \varepsilon < x < b + \varepsilon$.
7. Suppose first that $|z - z_0| \leq 1$, so that $z = z_0 + h$ with $|h| \leq 1$. Then $|z^n - z_0^n| \leq$ (something fixed)$\cdot|h|, \ldots$. So, given $\varepsilon > 0$, take δ to be ≤ 1 *and* $\leq$ some expression depending on ε.

Hints for Some Exercises

8. Since $|z^n| = |z|^n$, it is enough to show that $0 < r < 1$ implies $r^n \to 0$. But $1/r = 1 + h$ with $h > 0$, so $1/r^n \to +\infty$.
9. (a) Multiply and divide by $\sqrt{n^2 + 2n} + n$. (b) How does n compare to 2^n or to $\frac{1}{2}3^n$ if n is large? (c) In the product $n!$, about half the factors are $\geq n/2$.
11. See Section 3G.
12. (a) Divide the numerator and denominator of the fraction by n^2 and use (34). (b) Take the logarithms and view them as Riemann sums approximating a certain integral.
13. Is the sequence monotone and bounded? The limit is one solution of a quadratic equation.
14. The even and odd terms are monotone. The limit is a solution of a quadratic.
15. Look at $x_{k+1} - x_k$ and note that $x_n - x_0$ is a sum of such differences.
16, 17. If there is a limit, what equation must it satisfy? How do successive terms of the sequences relate to each other and/or the possible limits?

Section 3C

1. Show that $\{b_n\}$ is a Cauchy sequence.

Section 3D

3, 4. Write $w_n - z$ as a sum of n terms and note that, for large m, most of the terms are small by comparison with $1/n$. (Given $\varepsilon > 0$, choose N ....)
5. Adapt the method of Exercises 3 and 4.

Section 3E

2. There is a way, in principle, to construct such a subsequence.
4. Consider three possibilities: (i) there are infinitely many n such that $x_n > b$; (ii) there are infinitely many n such that $x_n = b$; (iii) neither of the first two.

Section 3F

4. All except the extreme cases can be deduced from Exercise 3 of Section 3D by taking logarithms.
5. Adapt the method of Exercises 3–5 of Section 3D.
9. What can you say about x_n/n for $n = mp$, where p is fixed and $m = 1, 2, \ldots$? Consider $\liminf(x_n/n)$.

Section 3G

1. Let $a_n = N^{1/n} - 1$ and consider (34).

Section 4A

1. Relate s_{2n-1} to s_{2n}.
2. Compute a few partial sums and use induction.
3. If $na_n \geq \varepsilon > 0$, then $s_n \geq \varepsilon$. Show by using similar inequalities that if $\{a_n\}$ is positive and nondecreasing and $\limsup(na_n) > 0$, then $\sum a_n = \infty$.

Section 4B

1, 2, 3. The Ratio Test.
4. Find a simple estimate for the size of the terms for large n. Show $a_n \geq 1/2n$ for large n.
5. Rationalize the numerator.
6, 7. The Ratio Test; (34) of Chapter 3.
8, 9. Cancel factors that are larger than 1 and compare to $\sum(1/n)$.
10, 11. The 2^m Test.
12. Put the two summands over a common denominator and use (32) of Chapter 3 to compare the series to $\sum(1/n\log n)$.
13. Use (6) of Chapter 3.
14, 15. See Exercise 1 of Section 3G.
16, 17, 18. Use Exercise 2 at the end of Chapter 4 *or* estimate the size of a_n by taking the logarithm and using (32) of Chapter 3. (To use the second method for 17 and 18, divide each factor by n.)
19. The 2^m test.
20–23. Group terms and estimate sizes.
26. Estimate how many such integers there are between 10^n and 10^{n+1} and get a (crude) upper bound for their contribution to the sum.

Section 4C

1. See Section 3G.

Additional Exercises for Chapter 4

1. Use (34) from Chapter 3 to deduce that $(n+1)^b/n^b \sim e^{b/n} \sim 1 + b/n$, so that the limit is b. *Or* use L'Hôpital's rule.
2. Convergence: The condition implies that $n(-1 + a_n/a_{n+1}) \geq c > 1$ for large enough n. Use Exercise 1 (with $1 < b < c$) and a comparison. *Or* show that, for large enough n, and for some $\varepsilon > 0$,

$$na_n - (n+1)a_{n+1} \geq \varepsilon\, a_n;$$

sum these inequalities to bound the partial sums of the series.
Divergence: The condition implies that, for large enough n, $a_{n+1}/a_n \geq [1/(n+1)]/[1/n]$, so compare to $\sum(1/n)$.

3. Use Exercise 3 of Section 4D.
4. Convergence: Look at $(n-1)\log(n-1)a_{n-1} - n\log na_n$. Divergence: Compare a_{n+1}/a_n to b_{n+1}/b_n with $b_n = 1/(n-1)\log(n-1)$.

Section 5A

1, 2, 3. Start with the Root Test.
4–8. The Ratio Test.
10. Choose $r < s < R$. What does the convergence of $\sum_0^\infty a_n s^n$ tell us about the size of $|a_n|$? and thus about convergence of $\sum_0^\infty |a_n|r^n$? Compare the modulus of a partial sum of $\sum_0^\infty a_n z^n$ to this series.

Section 5B

1. Let $w = 1 + z$.
2. Differentiate $1/(1-z)$.
3. Integrate twice.
4. Differentiate $1/(1-z)$ repeatedly.
5. Compute the coefficients b_n by using the Binomial Theorem and estimate them using Exercise 4.
6. (a) Go to the definition of the derivative.
7. Find equations that link any three successive coefficients. What are the first two coefficients?

Section 5C

4. Show that the partial sums satisfy $C_0 + C_1 + \cdots + C_n = A_0 B_n + A_1 B_{n-1} + \cdots + A_n B_0$; divide both sides by $n+1$ and use Exercises 3 and 5 of Section 3D.

Section 5E

2. Set $x_n = 1 - 1/n$ and consider

$$\sum_{k=0}^n a_k - f(x_n) = \sum_{k=0}^n a_k(1 - x_n^k) - \sum_{k=n+1}^\infty a_k x_n^k;$$

use the 2^m Test.

Section 6A

2. The trick is to prove the triangle inequality. For the second part, note that $\hat{d}$ distances are always < 1.

Section 6B

1–6, 10. Use the definitions.
7. The least upper bound is a limit point, if it is not in A.
9. Consider unions of sets like those in (1), e.g., the translate of these sets; you may also need something like the single point $\{0\}$. It is possible to construct A so that the seven sets are distinct.
11. What happens with the discrete metric? With $\mathbb{Q}$?
12. A may be chosen to be an interval.
13. For the first part, use the Least Upper Bound Property. Reduce the second part to (something like) the first part.

Section 6C

1, 2. Use the definition.

Section 6D

2. Given an open cover, some set of the cover contains p. How many points p_n are not in this set?
3, 5. Use the definitions.
6. Use a sequence ε_n converging to 0.
10. Prove the two parts separately, from the definitions.
11. Use the (extended) triangle inequality.
12. Choose $p_1 \in S$ and define a sequence with $p_{n+1} = f(p_n)$. Use the preceding exercise to show that this is a Cauchy sequence. For full details, see Section 15E.
13. Use Exercise 12.

Section 7A

1. Use the definitions.
2. $f^{-1}(B^c) = [f^{-1}(B)]^c$.
3, 4. Use the definitions.
5, 6. Use Exercise 4.
7, 8. Use Exercise 4 and take a reciprocal.
9. The image can be a half-open interval, like $(0, 1]$.
10. B can be the intersection of a real interval with $\mathbb{Q}$.
11. Use the definitions.
12. Use Exercise 11.
13. Use the definitions.
14. (a) Use the definitions. (b) How should g be defined at points of the closure that are not in A?
15. The proof of differentiability (Theorem 5.4) can be modified and simplified. *Or* show that the partial sums converge uniformly (see Exercise 3(b) of Section 5A).

Section 7B

1. Prove this for integers, then for rational x, and then use continuity.
2. Let $a = f(1)$. By induction, $f(n) = an^2$ for $n \in \mathbf{N}$. Also, $f(2x) = 4f(x)$, so $f(x) = ax^2$ for a dense set in $\mathbf{R}$.
3. Use the geometric meaning of the condition: The graph lies below any secant.
5. Prove the convexity condition for t of the form $p/2^n$ and use continuity.
6. Use differentiability; show that the partial sums converge uniformly on each subset $\{z : |z| \leq r\}$ for $r < R$.

Section 7C

1. Estimate the value of f_n at the point where it attains its maximum, or, for (d), where the denominator is minimal.
4. Evaluate at enough points to get control of the coefficients.
5. Use uniform continuity and connect the dots.
6. (b) Use the preceding exercise.
7. (a) Functions on $[0, 1]$ can be extended to be even functions ($f(-x) = f(x)$) on $[-1, 1]$. If polynomials converge uniformly to a function on $[-1, 1]$, what about their even part (terms of even degree)?
8. The P_n's are even, so consider the interval $[0, 1]$ and prove that $x - P_n(x) \geq 0$; then P_n increases with n.

Section 8A

1. (a)–(e) Remember that there are two versions of l'Hôpital's rule. (f) Consider h as the variable; use l'Hôpital twice.
3. (a) For a given $\varepsilon > 0$, this essentially reduces to the case of a bounded interval. (b) Take advantage of (a).
4. (b) Use part (a), and modify the construction of a function that is known to be discontinuous at every point (Example 2 in 8C).
5. Can f take the same value twice? Can it decrease, then increase (or vice versa)?
6. Reduce to the case $c = 0$ and use the preceding exercise.
7. The derivative at $x = 0$ is 0: The positive and negative contributions to the integral very nearly cancel – like looking at the tail of an alternating series.

Section 8B

1. After differentiating, understand the behavior of the numerator by looking at *its* derivatives.
2. IVT and monotonicity; to show that the right side gets large as b does, notice that $x \geq 1$ implies $1 + x^3 \leq 2x^3$.
3. (a) Use the IVT over a sufficiently large interval.
4. Integrate the polynomial and use the Mean Value Theorem.
5. Use the IVT and the noncountability of the (nonrational) reals in an interval.

Section 8C

1. The Intermediate Value Theorem; the Mean Value Theorem.
2. Although this function is not continuous on $[0, 1]$, it is uniformly continuous on $[\varepsilon/4, 1]$. Let $x_1 = \varepsilon/4$ and then deal with the rest of the partition.
3. If $f(c) > 0$, then $f(x) \geq \frac{1}{2}f(c)$ on some interval, so there is P such that $L(f, P) > 0$.
4. We need P with $U(f, P) < \varepsilon$. Take $x_1 = \frac{1}{2}\varepsilon$; then there are only finitely many bad points left to worry about.
5, 7. In any given bounded interval, for any given $\varepsilon > 0$ there are only finitely many points x where $f(x) \geq \varepsilon$.
6. The only candidates for points of differentiability are points of continuity. Look at difference quotients where one point is rational.

Section 8E

1. L'Hôpital; series expansion of e^x.
2. What general form do derivatives have for $x > 0$? Now use the preceding exercise.
3. Use (36) with $n = 2$ to conclude the following, and choose the best a:

$$|f'(x)| \leq 2a \cdot \sup |f| + \frac{1}{2a} \sup |f''|.$$

Additional Exercises for Chapter 8

1. What does the Mean Value Theorem say about f'?
2. Reduce this to the preceding exercise.
3. Approximate the integral with sums.
4. Use uniform continuity.
5. In the difference quotient $[L(x + h) - L(x)]/h$ write $x + h = x(1 + h/x)$ and use (i), then (ii).
6. Write $L(x + y) - L(x)$ as an integral and change variables so that the interval $[x, xy]$ becomes $[1.y]$.
8. Take $f_n \equiv 0$ outside $[1/2n, 1/n]$.
9. Do this first under the assumption that h is *uniformly* continuous. Then note that if we fix f and look at nearby functions g, we only need to know h on a bounded interval.
11. Use Exercise 7 and the Weierstrass Theorem to see that $\int_a^b f(x)^2 \, dx = 0$; then use Exercise 3 of Section 8C.

Section 9A

1. Look at e^{z_1}/e^{z_2}.
2. Start with (b) and use Exercise 5 of Section 2D.

3. See the preceding exercise.
4, 5. Use the definitions and the properties of the exponential function.
6. Fix $x \neq 0$. What does the n-th term look like if n is very large?
7. Let $u = e^z$ and note that we get a quadratic equation for u.
8. Use the identities in (11): The sum on the left becomes a pair of sums of geometric progressions.
12. Differentiate and use Exercise 11(a).
13. (b) Square the integral (using x and y to denote the two variables of integration), interpret the result as an integral in the plane, and change to polar coordinates.
15. Let $t = s/(1+s)$.

Section 9C

2. Use the fact that $|\tan x| \geq |x|$ on $(-\frac{\pi}{2} + \frac{\pi}{2})$ to show that $x^2/[2m\tan(k\pi/2m)]^2 \leq x^2/k^2\pi^2$.

Section 10B

1. Let m^+ be defined using closed intervals. It is easy to see that $m^+(A) \leq m^*(A)$. To prove the converse, replace any closed interval by a slightly larger open interval.
6. The union of all open intervals $I \subset A$ containing a point $x \in A$ is an open interval, so A is a union of disjoint open intervals. Any nonempty open interval contains a rational.
7. If there are only finitely many nonempty intervals, use (4) and induction. The finite case gives an inequality from which the general case follows.
10. Alter the construction of C somewhat and take advantage of Exercise 2.
11. The unbounded case can be reduced to countably many bounded cases.

Section 10C

1, 2. Use the definition.
3. Do this first for intervals with endpoints of the form $p/2^n$, p an integer.

Section 10D

2. Use (5).
5. It is enough to prove that some subsequence has this property. Passing to a suitable subsequence and renumbering, one may assume that $d(A_n, A_{n+1}) < 2^{-n-1}$. Then let $B_n = \bigcap_{m=n}^{\infty} A_m$ and $A = \bigcup_{n=1}^{\infty} B_n = \liminf A_n$. Show that $d(A_n, B_n) < 2^{-n}$ and $d(B_n, A) \to 0$.

Section 11A

4. (a) Prove this when A is a rectangle with sides parallel to the coordinate axes; show that any open set is a countable union of such rectangles.
6. Let

$$E_{kn} = \left\{ x \in A : \sup_{m \geq n} |f_m(x) - f(x)| > 1/k \right\}, \qquad k = 1, 2, \ldots, \quad n = 1, 2, \ldots.$$

Show that for each k there is $n = n(k)$ such that $m(E_{kn} < \varepsilon/2^{k+1}$, and show that convergence is uniform on the complement in A of the union of the $E_{k,n(k)}$.

Section 11D

7. $a\, 1_{E_a} \leq f$.
10. Use Exercise 7.

Section 11E

1. (c) For $t \geq 0$, we have $\log(1+t) \leq t$.
2. Apply the DCT to $f\, 1_{E_n}$.
3. (a) Use the DCT. (b) F has limits at $\pm\infty$.
4. Look at the sets where $f(x) > 1$ and where $f(x) \leq 1$.
8. The set $\{f > a\}$ is the union of the sets $\{f_n > a\}$; use the continuity property to estimate its measure.
9. Adapt the construction in the example after Theorem 12.5.

Section 12C

3, 4. It is enough to prove these for a dense subset. (Why?)
5. Use Theorem 12.7, and the method used in the proof of Theorem 12.5, to find a sequence of continuous functions that converge pointwise a.e., show that the limit is f a.e., and use Egorov's Theorem (Exercise 6 of Section 11A).
6. Show that for large enough a, the set $A = \{|f| > a\}$ has measure less than $\varepsilon/2$. Then, for any $n = 1, 2, 3, \ldots$, f is integrable on $[-n, n] \cap A$, and use can be made of Exercise 5.

Section 12D

9. Use Exercise 6, with $n = 2$.
10, 11. Cauchy-Schwarz.
12. Show that it is possible to choose a sequence of functions $h_n \in L^2$ such that $f h_n$ increases to f^2 and use the Monotone Convergence Theorem.

Section 12E

2, 3. Use Theorem 12.10. For Exercise 3, note that $m(A \cap B) = \int_B 1_A$.

Additional Exercises for Chapter 12

1. (a) It can be assumed that $b = 1$; the problem can be rephrased as a problem about a minimum value.
2. Replace f by f/a and g by ag in Exercise 1(b) and make a good choice of the constant a.
3. What choice of g would guarantee equality?
4. Use Exercises 2 and 3.
5. There are examples in which f and g are indicator functions.
6. Try functions that look like $x^a (\log x)^b$ for small positive x and like $x^c (\log x)^d$ for large positive x.

Section 13B

3. Use Exercise 2(a) and take a limit.
5. (a), (b) Use the expansion of $1/(1-z)$.
6. (b) Either approximate by continuous functions or break the interval $[b, 1)$ into at most two subintervals, each of which is obtained by translating by an integer a corresponding subinterval of $[-\pi, \pi)$.
7. Use (9), orthonormality.
8. Integrate by parts.
9. Use (43).

Section 13C

1–4. Dirichlet's Theorem is applicable. For 1(b), break the sum into even and odd terms.

Section 13D

1. Use Exercise 2 of Section 13B and Fejér's Theorem.
2. Use Fejér's Theorem.

Section 13E

1. Note that $\frac{1}{2\pi} \int_{-\pi}^{\pi} fg = 0$ for every trigonometric polynomial g. By the Weierstrass Approximation Theorem (or Fejér's Theorem), this implies that $\frac{1}{2\pi} \int_{-\pi}^{\pi} fg = 0$ for every continuous periodic g. Using continuous g's to approximate the indicator function of an interval $A \subset (-\pi, \pi)$, show that $\int_A f = 0$ for every such interval. Deduce from this and dominated convergence that $\int_A f = 0$ for every open set $A \subset (-\pi, \pi)$.

This and the approximation property of measurable sets imply that $\int_A f = 0$ for every measurable set. In particular, this is true for the sets $A = \{x : \operatorname{Re} f(x) > 0\}$ and $B = \{x : \operatorname{Re} f(x) < 0\}$. Therefore

$$\frac{1}{2\pi} \int_{-\pi}^{\pi} |\operatorname{Re} f| = \int_A \operatorname{Re} f - \int_B \operatorname{Re} f = 0,$$

which implies that $\operatorname{Re} f = 0$ a.e. The same argument applies to $\operatorname{Im} f$.

2. Use the preceding exercise.

Section 13F

1. If g belongs to this space, then $f - S_N f$ is orthogonal to $S_N f - g$, which implies that $\|f - g\|$ is at least equal to $\|f - S_N f\|$ (see Exercise 9 of Section 12D).
2, 3. Cauchy-Schwarz.

Section 13G

1, 2. Use Theorems 13.13 and 13.14 and Proposition 13.4.
6. (a) Use Lemma 13.15 and Theorem 13.14 to conclude that $S_N f$ converges uniformly to a continuous function g and note that g has the same Fourier coefficients as f.
8. Proposition 13.4, iterated.
9. Use Theorem 13.14 and Proposition 13.4.
10. Translate f, so you may assume that $x_0 = 0$. Then write f as a sum of a function to which Dirichlet's Theorem applies and a multiple of the function which is -1 on $(-\pi, 0)$ and $+1$ on $(0, \pi)$.

Section 14A

1. See the hint for Exercise 10 of Section 13G.
2. Show that the values of $T_N f(x)$ lie in the interval $(-1, 1)$.

Section 14B

1. Write $f(x) - f(y)$ as a series; estimate the sum of the terms up to $2^{-n} \sim |x - y|$ by using the size of the coefficients and an estimate on the difference of the cosine terms; estimate the remaining sum using only the size of the coefficients and the boundedness of the cosine function.

Section 14D

2. Note that $\int_0^x g = G(x)$ is even and periodic. Solve for f_1 and f_2 in terms of f and G.

Sections 14E and 14F

Expand the would-be solution in appropriate eigenfunctions and proceed in analogy with the vibrating string problem.

Section 14H

1. Use the Dominated Convergence Theorem.
2. Integrate by parts.
6. Note that f satisfies the differential equation $df/dx = -xf(x)$. Show that $\widehat{f}$ satisfies the same equation (with respect to the ξ variable), by using the results of the preceding exercises.
   Deduce that $\widehat{f}$ is a multiple of $e^{-\xi^2/2}$. Because f has integral $= 1$, it follows that $\widehat{f}(0) = 1$.
7. Use the inversion formula (43) and the results of previous exercises.
8. Show that the Fourier coefficients $\widehat{g}$ of the periodic function g on the right side are related to the Fourier transform of f by

$$\widehat{g}(n) = \frac{1}{2\pi} \cdot \widehat{f}(m)$$

and use Theorem 13.16.
9. It is enough to show that $\widehat{f}$ is determined by these values. But consider $\widehat{f}$ as a function on the interval $[-M/2, M/2]$ and look at its Fourier expansion in terms of the complex exponentials that are periodic with period M, just as was done above for f_L in (37).

Section 14I

3. When does equality hold in the Schwarz inequality?

Section 15A

3. Write this as $f(x)e^{-Cx} - f(0) \leq 0$.

Section 15B

4. Use Exercise 2 to find one solution and Theorem 15.1 to find the rest.
8. The argument will be different depending on whether r is a root of p.
10. (a) The differential equation and the boundedness condition determine G on the intervals $(-\infty, 0)$ and $(0, \infty)$ up to constant multiples. The remaining conditions determine the constants.
12. Given a linear combination that vanishes, look at its behavior for large x.

Section 15C

6. Differentiate the inner product $(u(x), u(x))$. (This can be done either directly from difference quotients and properties of the inner product or by writing out the inner product.)
8. Differentiate $e^{xA+B}e^{-xA}e^{-B}$ and compare the values at $x = 0$ and $x = 1$; remember that $e^A e^{-A} = I$.
10. Use implicit differentiation.
11. Use Exercise 10.
12. Note that $f(x, x^a v) = x^{a-1} f(1, v)$.

Section 15E

5. Consider the presence or absence of a Lipschitz condition.
8. Induction.

Section 15F

4. What is the dimension of the space of solutions?

Notation Index

A^c, 77
$A \triangle B$, 14
$\mathbb{C}$, 26
$C(A; \mathbb{C})$, $C(A; \mathbb{R})$, 91
$C^\infty(\mathbb{R})$, 219
D_N, 182
$f \vee g, f \wedge g$, 146
$f^{-1}(A)$, 87
$\widehat{f}(n)$, 176;
$\widehat{f}(\xi)$, 211
F_N, 184
inf, 16
lim inf, lim sup, 34
L^1, 162
L^2, 166

L^p 172
$L^2(I)$, 175
$L(f, P)$, 107
$N_\varepsilon(p)$, 75
$\mathbb{N}$, 1
1_A 149
$P(D)$, 219
$\mathbb{Q}$, 2
$\mathbb{R}$ 12,15
S_N, 179
sup, 16
T_N, 184
$U(f, P)$, 107
$\mathbb{Z}$, 2

General Index

Abel's Theorem, 71
absolute convergence, 46
additive function, 91
a.e. (almost everywhere), 158
algebraic number, 24
almost everywhere (a.e.), 158
alternating series, 55
approximate identity, 195
Archimedean property, 8
Ascoli-Arzelà Theorem, 227
axioms
    for $\mathbb{N}$, 5
    for $\mathbb{Q}$, 5, 7, 8
    for $\mathbb{R}$, 15–16
    for $\mathbb{Z}$, 5, 7

Banach Fixed Point Theorem, 84, 231
Banach-Tarski Theorem, 132, 239
band-limited signal, 215
Bernstein polynomials, 95
Bessel's inequality, 179
Beta function, 129
binary expansion, 22
Bolzano-Weierstrass Theorem, 39
boundary conditions, 204, 208
boundary of set, 79
bounded sequence, 30
    set, 80
    function, 88

Cantor set, 84, 134
Cauchy product, 67

Cauchy sequence, 35
Cauchy-Schwarz inequality, 166, 168
chain rule, 100
closed set, 76
closure of set, 78
commutative group, 7
commutative ring, 7
compact set, 79
compactly supported function, 164
comparison test, 48
complete metric space, 81
complex conjugate, 27
complex numbers, 26
composition of functions, 87
conditional convergence, 54
connected set, 79
continuity, 86
    uniform, 88
Contraction Mapping Theorem, 84, 231
convergence
    of sequences, 30, 81
    of series, 45
convergence tests
    alternating series, 55
    comparison, 48
    Gauss, 60
    integral, 52
    Raabe, 59
    ratio, 48
    root, 49
    2^m, 50

convergent sequence, 30, 81
  series, 45
convex function, 91
convolution, 195
cosh, 123
cosine, 122
countable additivity, 139
countable set, 23
cover, 79
  open, 79
cut, 10

DCT (Dominated Convergence Theorem), 155
decimal expansion, 22
dense subset, 93
derivative, 64,
differentiable, 99
differentiation, 99
  of power series, 64
diffusion equation, 207, 214
Dirichlet kernel D_N, 182
Dirichlet's Theorem, 183
discrete metric, 73
disk of convergence, 63
divergent series, 45
Dominated Convergence Theorem (DCT), 155

Egorov's Theorem, 147
eigenfunctions, 204
equicontinuity, 227
equidistributed sequence, 202
equivalent metrics, 75
Euler's constant, 57
Euler's formula for $\sin x$, 125
exponential function, 69, 106, 119–122
extended reals, 40

fast Fourier transform (FFT), 210
Fatou's Lemma, 156
Fejér kernel F_N, 184, 185
Fejér's Theorem, 186
FFT (fast Fourier transform), 210
Fibonacci sequence, 44
field, 18

Fourier
  coefficients, 176, 177
  series, 176
  transform, 212
function, 86
  additive, 91
  Beta, 129
  bounded, 88
  compactly supported, 164
  continuous, 86
  continuous at a point, 86
  convex, 91
  differentiable, 99
  Gamma, 129
  hypergeometric, 130
  indicator, 149
  integrable, 152, 165
  integrable periodic, 175
  integrable simple, 149
  inverse, 105
  L^2-periodic, 175
  measurable, 152
  monotone, 102
  nondecreasing, 102
  nonincreasing, 102
  periodic, 173
  square-integrable periodic, 175
  strictly decreasing, 102
  strictly increasing, 102
  strictly monotone, 102
  uniformly continuous, 88
fundamental solution, 223
Fundamental Theorem
  of Algebra, 124
  of Calculus, 111

Gamma function, 129
Gauss's test, 60
Generalized Mean Value Theorem (GMVT), 102
geometric series, 46
Gibbs phenomenon, 197
GMVT (Generalized Mean Value Theorem), 102
greatest lower bound, 16
Greatest Lower Bound Property, 16
Gronwall's inequality, 219

Hardy-Littlewood
  inequality, 169
  maximal function, 169
heat equation, 207, 214
Heine-Borel Theorem, 81
Heisenberg Uncertainty Principle, 217
Hilbert space, 166
Hölder's inequality, 172
hyperbolic functions, 123
hypergeometric function, 130

imaginary part, 27, 165
indicator function, 149
inequality
  Bessel's, 179
  Cauchy-Schwarz, 166, 168
  Gronwall's, 219
  Hardy–Littlewood, 169
  Hölder's, 172
  isoperimetric, 200
  triangle, 73
infimum, 16
infinite product, 126
initial conditions, 204
integers 2
  construction 6
integrable function
  Lebesgue, 152, 165
  Riemann, 109, 110
integrable simple function (ISF), 149
integral (Lebesgue)
  of integrable function, 152
  of ISF, 149
  of nonnegative function, 151
integral (Riemann), 109
integral test, 52
integration by parts, 115
interior point, 75
interior of set, 76
Intermediate Value Theorem, 105
inverse function, 105
inverse image of set, 87
ISF (integrable simple function), 149
isoperimetric inequality, 200

Jensen's inequality, 117

least upper bound, 16
Least Upper Bound Property, 16
Lebesgue's theorem on
  dominated convergence, 155
  Fourier coefficients, 193
L'Hôpital's Rule, 103
limit
  lower, 33
  of sequence, 30
  point, 76
  upper, 33
Lipshitz condition, 231
logarithm (natural), 42, 102
lower
  bound, 16
  integral, 108
  limit, 33
  sum, 107
Lusin's Theorem, 166

Mandelbrot set, 44
maximal function, 169
Mean Value Theorem (MVT), 101
measurable
  function, 145
  set, 136
Mertens' Theorem, 67
method of successive approximations, 230
metric, 73
  discrete, 73
  standard, 74
metric space, 73
  compact, 79
  complete, 81
  separable, 94
metrics, equivalent, 75
modulus, 27
momentum operator, 217
Monotone Convergence Theorem, 156
monotone function, 102
  sequence, 32
MVT (Mean Value Theorem), 101

natural numbers, 1
neighborhood, 75
Nested Interval Property, 20
no-gap property, 4, 32

nondecreasing
   function, 102
   sequence, 31
nonincreasing
   function, 102
   sequence, 32
nonmeasurable set, 142
norm, 74
null set, 158

open cover, 79
open set, 75
ordered field, 18
orthogonal, 168
orthonormal, 168, 174
outer measure, 133

Parseval's Identity, 190
partial sums, 45
partition, 107
   refinement, 108
Peano's Theorem, 228
periodic
   function, 173
   integrable function, 177
   square-integrable function, 175
π, 121
Picard iterates, 232
Picard's method, 230
Plancherel's Theorem, 212
point
   interior, 75
   limit, 76
point of density, 171
Poisson summation formula, 215
polar decomposition, 28
position operator, 217
positive number, 17
power series, 61
   differentiation, 63
   radius of convergence, 61
product
   of power series, 68
   of series, 67
proof, discussion, 12–13
proper subset, 79

Raabe's test, 59
radius of convergence, 61
ratio test, 48
rational numbers, 2
   construction, 7–8
real numbers
   axioms, 15–16
   construction, 10–12
   extended, 40
real part, 27, 165
rearrangement (of series), 55
refinement of partition, 108
Riemann integrable, 109
Riemann sum, 113
Riemann's theorem on rearrangement, 56
Riemann-Lebesgue Lemma, 178
Riesz-Fischer Theorem, 189
Rolle's Theorem, 101
root test, 49

Schrödinger equation, 217
separable differential equation, 226
separable metric space, 94
sequence, 30
   bounded, 30, 82
   Cauchy, 35, 82
   convergent, 30, 81
   Fibonacci, 44
   monotone, 32
   nondecreasing, 31
   nonincreasing, 32
sequences
   difference, 37
   product, 37
   quotient, 37
   sum, 37
sequential compactness, 83
series, 45
   absolutely convergent, 46
   alternating, 55
   conditionally convergent, 54
   convergent, 45
   divergent, 45
   geometric, 46
set
   bounded, 80

bounded above, 16
bounded below, 16
Cantor, 84
closed, 76
closure of, 78
compact, 79
connected, 79
interior of, 75
measurable, 136
null, 158
open, 75
sequentially compact, 83
sine, 122
sinh, 123
spectral theorem, 175
square-integrable periodic function, 175
standard metric, 74
standing-wave solutions, 205
strict contraction, 84, 231
strictly
  decreasing, 102
  increasing, 102
  monotone, 102
subcover, 79
subsequence, 39
subset
  compact, 79
  dense, 93
  proper, 79
sum of series, 45
supremum, 16
symmetric difference, 142

Tauberian theorem, 72
Taylor polynomial, 114
Taylor's formula
  with remainder, 115
  with integral remainder, 116
ternary expansion, 22
Theorem
  Abel, 71
  Ascoli-Arzelà, 227

Banach, 84, 231
Banach-Tarski, 132, 239
Bolzano-Weierstrass, 39
Dirichlet, 183
Egorov, 147
Fejér, 186
Heine-Borel, 81
Liouville, 25
Lebesgue, 155, 193
Lusin, 166
Mertens, 67
Peano, 228
Plancherel, 212
Riesz-Fischer, 189
Weierstrass, 93, 187, 199
Riemann, 56
Rolle, 101
Weyl, 202
transcendental number, 24
translate, of set, 134
translation invariance, 139
traveling-wave solutions, 206
triangle inequality, 73
trigonometric functions, 122
trigonometric polynomial, 187
2^m test

uncountable set, 23
uniform continuity, 88
uniform convergence, 93
upper
  bound, 16
  integral, 108
  limit, 33
  sum, 107

wave function, 216
Weierstrass's theorem on
  trigonometric approximation, 187
  nondifferentiable function, 199
  polynomial approximation, 93
Weyl's theorem on equidistribution, 202